SpringerBriefs in Environment, Security, Development and Peace

Volume 30

Series editor

Hans Günter Brauch, Mosbach, Germany

More information about this series at http://www.springer.com/series/10357
http://www.afes-press-books.de/html/SpringerBriefs_ESDP.htm
http://www.afes-press-books.de/html/SpringerBriefs_ESDP_30-31.htm

Bin Su · Elspeth Thomson
Editors

China's Energy Efficiency and Conservation

Sectoral Analysis

Springer

NUS National University of Singapore
Energy Studies Institute

Editors
Bin Su
Energy Studies Institute
National University of Singapore
Singapore
Singapore

Elspeth Thomson
Victoria, BC
Canada

ISSN 2193-3162 ISSN 2193-3170 (electronic)
SpringerBriefs in Environment, Security, Development and Peace
ISBN 978-981-10-0735-4 ISBN 978-981-10-0737-8 (eBook)
DOI 10.1007/978-981-10-0737-8

Library of Congress Control Number: 2016934434

Cover map: This map is from the Perry-Castañeda Library Map Collection. Used by permission of the University of Texas Libraries, The University of Texas at Austin at: https://www.lib.utexas.edu/maps/cia14/china_sm_2014.gif. More on this book is at: http://www.afes-press-books.de/html/SpringerBriefs_ESDP_30-31.htm.

Copyediting: PD Dr. Hans Günter Brauch, AFES-PRESS e.V., Mosbach, Germany

Printed on acid-free paper

This Springer imprint is published by Springer Nature
The registered company is Springer Science+Business Media Singapore Pte Ltd.

Acknowledgements

The Energy Studies Institute (ESI) at the National University of Singapore started its China energy research in 2012. Its first conference, "China Energy Issues in the 12th Five-Year Plan and Beyond," held in February 2012, examined the economic, environmental, and security aspects of China's energy and carbon mitigation strategies. During this event, speakers and participants shared their opinions on China's overall energy developments, and some of these discussions have been published in journals, such as in volume 73 of *Energy Policy* (Special Issue, 2014).

ESI has established formal relationships with energy think-tanks in China—such as the Institute of Policy and Management, the Chinese Academy of Sciences, and the College of Economics and Management in Nanjing University of Aeronautics and Astronautics—to look into China's latest energy issues and their influences on the region. In 2013, ESI launched a series called the "Singapore–China Energy Forum" to discuss the opportunities and challenges faced by China's recent and future energy developments. The topics included energy efficiency and conservation, energy and carbon markets, energy security, climate change, and many others. This volume is the compilation of some of the presentations on the subject of energy efficiency and conservation delivered at the first forum, held in November 2013.

On behalf of the ESI research team, I would like to express my sincerest thanks to our Executive Director, Prof. S.K. Chou, and ESI's board members for their unwavering support for our China energy research and the Singapore–China Energy Forum series. We are also grateful to our administrative colleagues, including Mr. Peter Yap, Ms. Jan Lui, and Ms. S. Telagavathy, who assisted in the management of our events. Our special thanks also go to ESI's Publications Committee, especially our Editor, Ms. Eunice Low, who spent much time reading and copy-editing drafts of our manuscripts. Last but not least, we would like to thank all the speakers, authors, reviewers, participants, and research partners.

With our continued efforts in China energy research, we hope to generate sustained and constructive discussions in this area.

Singapore
November 2015

Dr. Bin Su

Contents

Chapter 1
Introduction

Bin Su and Elspeth Thomson

China is the largest energy consumer and carbon emitter in the world. At the 2009 United Nations Climate Conference in Copenhagen, the Chinese government committed to reducing its carbon intensity (CO_2 emissions per unit of GDP) by 40–45 % by 2020 compared to 2005 levels. Then, in a US–China Joint Announcement on Climate Change, released on 11 November 2014, the government announced that it would reduce the growth of its CO_2 emissions such that they would peak around the year 2030. About seven months later, the government on 30 June 2015, in its "Intended Nationally Determined Contributions (INDCs)" submitted to the United Nations Framework Convention on Climate Change (UNFCCC), set a new target to reduce the country's carbon intensity by 60–65 % by 2030 compared to 2005 levels.

To achieve these reduction targets and sustainable development, the Chinese government is carefully re-examining its energy development strategies together with its long-term economic strategies. Besides the national reduction targets set in the 12th Five-Year Plan (FYP 2011–15), China's State Council further announced the "12th Five-Year Plan for Energy Development" in January 2013. This plan summarises the country's major achievements in energy development in the 11th FYP, and clarifies national energy policy priorities through an ambitious set of infrastructure and market targets in the 12th FYP. Energy efficiency improvement and conservation are important mechanisms for China's energy development, and are identified as the most effective ways to manage and restrain the growth in energy consumption and reduce global emissions.

Dr. Bin Su, Senior Fellow, Energy Studies Institute, National University of Singapore, Singapore; Email: subin@nus.edu.sg.

Dr. Elspeth Thomson, Consultant; Former Senior Fellow and Head of the Energy and Environment Division, Energy Studies Institute, National University of Singapore.

B. Su and E. Thomson (eds.), *China's Energy Efficiency and Conservation*,
SpringerBriefs in Environment, Security, Development and Peace 30,
DOI 10.1007/978-981-10-0737-8_1

China's Energy Statistical Yearbook 2013 indicates that China's total energy consumption had increased to 3.4 billion tonnes of coal equivalent in 2012, and its equivalent CO_2 emissions were estimated at 8.313 million tonnes. The largest emitting sector was the power sector (around 48.1 % of total CO_2); followed by the manufacturing sector (~36.6 %); transport sector (~7.1 %); building sector (~4.1 %); and household sector (~4.1 %).

Volume I of *China's Energy Efficiency and Conservation: Sectoral Analysis*, is a compilation of six of the presentations on China's energy efficiency and energy conservation that were delivered at the *1st Singapore–China Energy Forum* organised by the Energy Studies Institute at the National University of Singapore in November 2013. The remaining six presentations are found in Volume II. The 12 speakers for this event were invited from mainland China, Hong Kong, Taiwan, Australia, Singapore and Japan.

Volume I examines energy efficiency and conservation in five specific sectors: power generation, energy-intensive manufacturing, iron and steel, transport and building. Volume II discusses the energy consumption behaviour of Chinese households; estimates potential energy savings and GHG emission reductions resulting from various energy efficiency measures; calculates the total factor energy efficiency and pollution efficiency indices for the country's 29 provinces and regions; looks at China's basic energy conservation and energy efficiency legislation and policy frameworks; and compares China's approaches to energy efficiency with those of Japan.

Specifically, in Chap. 2 of this volume, Yuan Jiahai and his co-authors highlight how the power sector is usually responsible for the largest portion of national and regional carbon emissions. They review China's power sector developments over the past 35 years, and identify the key driving factors behind the success or failure of the energy efficiency improvements in the power sector, namely: laws and regulations; technical standards and management systems; and technology innovation and economic incentives. They also construct an energy efficiency scenario to study energy efficiency and conservation potentials up to 2020, and discuss how these potentials could be achieved.

Ma Chunbo, in Chap. 3, reviews the impacts of changing policy parameters, such as the unbundling reforms geared towards enhancing power sector efficiency, and provides analysis based on a large panel set of power plants (300–600) in China from 1997 to 2010. Ma applies data envelopment analysis models to evaluate the relative performance of various power plants, and examines technology mandates, such as the small-unit shutdown mandate (SUSM), and the "promoting the big and quashing the small policy" (PBQSP). These are accompanied by a discussion of the potential impact on power plant efficiency and an integrated analysis of the policy parameters and challenges faced in the regulation of power plants.

China has experienced intensive industrialisation and motorisation since the 2000s, and energy efficiency is expected to moderate the increasing demand for energy, especially in the industrial sectors. In Chap. 4, Wu Libo discusses the effectiveness of policies announced or implemented in recent years (1994–2010), which were designed to optimise energy conservation in China's energy-intensive manufacturing sectors. Based on the TIMES modelling of Shanghai's energy system, Wu analyses the projected technological development in various energy-intensive sectors up to 2050 under different scenario settings.

Fan Ying and her co-authors in Chap. 5 examine China's iron and steel industry, which is one of the largest carbon-emitting sectors in the world. They identify its major carbon abatement technologies, and forecast abatement potentials up to 2020/2030. By using abatement cost curves, the cost of carbon abatement in China's iron and steel industry can be estimated. This information can then be used to obtain the marginal abatement cost curve. They construct a two-country and two-goods partial equilibrium model to simulate the influence of carbon pricing and allowance allocation to the iron and steel industry.

Transport accounts for about 23 % of global CO_2 emissions and 7–8 % of national GHG emissions in China. To reduce the transport energy demand and GHG emissions in China, many policies have been considered, including implementing fuel economy standards, promoting advanced vehicles as well as alternative fuels and developing a high-speed railway system. Huo Hong, in Chap. 6, analyses previous and soon-to-be implemented policies in China aimed at controlling energy use and emissions from transport. She also examines challenges and opportunities for China in building a low-carbon, clean and sustainable transport system.

Building sectors account for approximately 40 % of global energy consumption and 20 % of China's total energy consumption. In Chap. 7, Wang Shengwei and Gao Dian-ce present the basic approaches to energy-saving in buildings, and introduce the concept of building life-cycle commissioning and optimisation, with particular reference to heating, ventilation and air conditioning (HVAC) systems. The methodologies and approaches used for commissioning and optimisation at different construction stages are summarised. Wang and Gao explain on-site implementation of the control strategies, and provide case studies involving energy optimisation in new buildings and energy performance assessment in existing buildings in Hong Kong.

A key message from the analysis of five sectors in this volume, is that to achieve the energy-saving and mitigation targets, more efforts are required in terms of: employing best practices in energy efficiency and conservation projects; identifying energy efficiency and conservation potentials; making advanced energy efficiency and conservation technologies available; and very crucially, improving the relevant regulatory frameworks. It is also important for the Chinese government to take into account its significant regional diversities, and to understand the behavioural changes that are needed across society if further energy efficiency improvements are to be realised.

With globalisation, countries around the world are highly connected through an international supply chain. China can benefit from the past experiences of developed countries such as Japan, the United States and EU nations. At the same time, China's experiences can also provide valuable lessons for other developing countries, such as India and the nations of Southeast Asia.

Reference

National Bureau of Statistics (2014) *China's Energy Statistical Yearbook 2013*. Beijing: China Statistics Press.

Chapter 2
Energy Efficiency and Conservation in China's Power Sector: Progress and Prospects

Jiahai Yuan, Chunning Na and Mian Yang

Abstract This chapter addresses the progress and prospective of energy efficiency (EE) and conservation in China's power sector. To better understand China's successes and failures in EE in the power sector, the institutional characteristics of the sector are first briefly analysed. Then key EE drivers and the achievements in the past years are summarised. An energy efficient scenario for the power sector is constructed to probe the EE potential into 2020. Energy conservation potential is estimated at more than 300 Mtce, accompanied with vast co-benefits of CO_2 and air pollutants abatement. Policy implications are proposed to fully deploy the potential in the sector.

Keywords Power sector · Energy efficiency · China

2.1 Introduction

China has witnessed spectacular development in its power sector following 35 years of reform and opening-up. However, worsening global climate change and sustainable development situations have posed pressing requirements to the sector. Meanwhile, opportunities are there to realise a frog-leap in improving energy efficiency.

This chapter addresses the progress and prospects of energy efficiency (EE) and conservation in the power sector. In literature, EE is a concept with multiple meanings. In terms of the power sector, its narrow definition applies to the heat rate of power generation, especially in thermal power plants. Technical progress as well

Dr. Jiahai Yuan, Associate Professor, School of Economics and Management, North China Electric Power University; Email: yuanjh126@126.com.

Ms. Chunning Na, Ph.D. Student, School of Economics and Management, North China Electric Power University; Ningxia Key Laboratory of Intelligent Sensing for Desert Information, School of Physics & Electrical Information Engineering, Ningxia University.

Dr. Mian Yang, Associate Professor, Economics and Management School, Wuhan University; Center of Population, Resources & Environmental Economics, Wuhan University.

B. Su and E. Thomson (eds.), *China's Energy Efficiency and Conservation*,
SpringerBriefs in Environment, Security, Development and Peace 30,
DOI 10.1007/978-981-10-0737-8_2

as innovation in the operation can improve generation efficiency. A wider concept applies to energy input of the net electricity service delivered to the customers. In other words, the energy loss due to own consumption of the power plants and line loss of the power system is deduced from the original power generation. In this sense, power grid infrastructure and its operation model have direct impact on EE in the sector. In a broader sense, EE in the sector also means gaining the same energy service with few unfavourable by-products, given that all kinds of power generation technologies, especially coal power, have certain levels of pollutions or other disagreeable effects. In this sense, EE in the power sector also implies improvement in the generation mix. And finally, whatever is consumed by households or producers, the purpose of electric power service is to gain certain kinds of 'utility' (i.e. cooking or lighting) or 'value' (i.e. profit). Therefore, in its broadest sense EE can be defined as the utility value via a certain amount of energy input. At this level, EE in the power sector should include the efficiency improvement of terminal electric equipment, optimised pattern of energy usage, or adjustment towards more value-added production/service activities. The former two can be realised by demand side management (DSM), while the last belongs to structural energy conservation—a topic beyond the scope of this chapter.

The remainder of the chapter is organised as follows. Section 2.2 briefly reviews the characteristics of China's power sector and its institutional arrangements, which largely determine the successes and failures of EE in the past. Section 2.3 discusses the EE drivers and measures. Section 2.4 presents the achievements of EE in the sector. Section 2.5 examines a baseline scenario of the power sector and an alternative EE scenario in the period 2012–20. Section 2.6 addresses challenges and policy implications. This chapter concludes with Sect. 2.7.

2.2 Characteristics of China's Power Sector

Since the birth of new People's Republic of China, the country has gradually established the Leninist planned economy, with energy regarded as the key driving force of economy growth. Therefore, supply capacity has always been the top priority of the power sector. Due to China's energy resource endowment, the fuel mix of power generation is dominated by coal, which in turn contributes to serious pollutant emissions.

State ownership is an integral part of planned economy and 35 years of reform and opening-up are largely a process of economic liberalisation, during which competition has been gradually introduced. Though China's power sector has experienced four stages of institutional reform, and grid and generation business were successfully separated in 2002, state-ownership is still the key component of the institutional arrangements in the sector. The power grid is 100 % owned by the state while more than 90 % of the generation assets are owned by central and local governments. State ownership may be beneficial to the implementation of command-and-control EE measures that the Chinese government is familiar with, but will pose challenges to market-based measures in the future.

Meanwhile, in the planned economy, price is distorted in that it reflects the intention of the government, instead of the scarcity of the resources. The prices of various energy products have long been kept at low levels to promote economic growth. Therefore, energy efficiency is notoriously poor in China's power sector.

Until recently, the only achievement of the 2002 reform is the separation of generation and grid business. Five national generators were created to foster expected competition; a small regional grid company (China Southern Grid) was established in parallel with State Grid Company of China, franchised with exclusive transmission, distribution and sales business within their business area. Though it was intended to introduce market-oriented price reform, the progress is rather the trivial: the wholesale power generation price is strictly under the control of the central government, while the retail price is approved by provincial governments.

2.3 EE Drivers in China's Power Sector

Ever since the 10th Five-Year-Plan (FYP) period (2001–05), the Chinese government has taken measures to deal with the resource-ecology-environment puzzle. For the power sector, the low-carbon development is essentially a radical innovation to the energy-economy nexus, not simply for GHG emissions control. In turn, it is concerned with the low-carbonisation of power generation, transmission, distribution, retail and utilisation, and depends very much on the drivers of the legal system, technical standards and innovations, and economic incentives. In this section, we discuss the main EE drivers since the 2002 sector reform, with an emphasis on the 11th FYP period (2006–10). Our analysis will also cover 2011–12, where relevant data are publically available.

2.3.1 Drivers of Law and Regulation

Since the new century and particularly in the 11th FYP period (2006–10), the Chinese government has issued a package of policies to promote clean energy development and reinforce the low-carbon pathway (Table 2.1). In particular, the Renewable Energy Law provides legal foundation for renewable energy development (NEA 2005), while the Energy Conservation Law highlights energy conservation as a basic national policy (NEA 2007).

China also pledged to cut its GDP CO_2 emissions intensity by 40–45 % relative to the 2005 levels by 2020 and formulated several policies to reduce pollutant emissions in the power sector (Table 2.2).

Table 2.1 Policies on power planning and power system operation in China

No.	Title		Time issued
1	Power planning and power system operation	The medium-and-long-term plan for renewable energy development	2007
2		The energy conservation and emissions reduction plan during the 12th FYP period	2012
3		The renewable energy development plan during the 12th FYP period	2012
4		The special plan on the industrialisation of key smart grids technologies	2012
5		Rules on forbidding the construction of generation units smaller than 135 MW	2002
6		Rules on the closure of small coal power generation units	2007
7		Rules on energy conservation dispatch (trial)	2007
8		Rules on the upgrading and retrofitting of coal power generation plants	2012

Source Compiled by the authors from various policy files issued by NDRC

Table 2.2 Emissions control policy in the power sector

No.	Title	Time issued
1	Rules on curbing SO_2 emissions from coal power plants	2003
2	Rules on flue gas desulphurisation in coal power plants	2007
3	Rules on the statistics and supervision of key pollutants emissions during the 12th FYP period	2013

Source Compiled by the authors with files issued by NDRC

2.3.2 Drivers of Technical Standard and Management System

Since 2011, the China Electricity Council (CEC) has been entrusted by the National Development and Reform Commission (NDRC) and the National Energy Administration (NEA) to implement EE benchmarking in SOE coal power plants. In 2012, the work program on the supervision of EE and conservation in the power sector (CEC 2012), as well as the supervision and evaluation measures on EE indexes in power enterprises (NEA 2012a) were put into operation to measure the energy efficiency levels in both power generation and grid enterprises. A series of technical standards, including the guide on clean production in coal power plants, the technical manual on the evaluation of energy consumption status in coal power plants, etc., were promulgated by NEA (2012a) to promote EE in the power sector.

2.3.3 Drivers of Technology Innovation

The key technical EE measures in the power generation side include: retrofitting of existing coal plants, clean coal generation technologies, combined heat and power generation (CHP) or advanced CCHP, and non-fossil generation technologies. For instance, in coal power generation, the EE potential of clean coal technology is vast because of China's heavy reliance on coal power. As is shown in Table 2.3, the heat rate of a 600 MW coal generation unit is 40 g standard coal equivalents (gce) less in ultra supercritical (USC) than in subcritical unit. Assume an annual operation of 5,500 h; a 600 MW USC unit will consume 56.8 thousand tons standard coal equivalents (tce) less than a subcritical unit. Therefore, technical innovation plays an important role in the EE of the sector.

On the power grid side, the development of smart grids has made significant contributions to EE improvement. Grid companies have grasped the engineering of UHV transmission systems and have developed cutting-edge AC (1,000 kV) and DC (±800 kV) transmission lines in China. A dozen new technologies, including flexible AC transmission systems (FACTS), intelligent substations, smart distribution technologies, intelligent meters and smart demand response, etc., are implemented or under construction in the power grids (Yuan et al. 2014a, b).

2.3.4 Drivers of Economic Incentive

Though mandatory legal and administrative measures play a major role, economic incentive is more important. As a matter of fact, administrative measure are always supplemented by economic incentives. For instance, the Chinese government provides the generators with economic compensation for closing small coal units, and rewards and favourable loans for retrofitting existing generation plants.

Table 2.3 Thermal efficiency and heat rate of various coal generation technologies

Technology	Steam temperature (°C)	Steam pressure (MPa)	Thermal efficiency (%)	Heat rate (gce/kWh)
Medium temperature and pressure	435	35	24	480
High temperature and pressure	500	90	33	390
Ultrahigh pressure	535	13	35	360
Subcritical	545	17	38	324
Supercritical	566	24	41	300
USC	600	27	43	284
700 °C USC	700	35	>46	210

Source CEC (2013b)

Table 2.4 Electricity price policy related with energy saving and emission reduction in China

Type	Policy	Time issued
Differential tariff	Policy on improving the differential tariff	2006
Tiered tariff	The guiding policy on implementing tiered tariff policy in household	2010
Renewable power price	Policy on renewable energy price and cost allocation	2006
	Policy on renewable energy surcharge	2007
	Policy on improving the generation price of wind power	2009
	Policy on the management of renewable energy fund from price surcharge	2012
Coal power price	Measures on desulphurisation coal generation price	2007
Punitive price for energy-intensive industries	Policy on forbidding the preferential price in energy-intensive industries	2010
Price on power generation right trading	Policy on regulating the price of power generation right trading	2009

Source Compiled by the authors from various document files issued by NDRC

The Pricing mechanism is the core of the economic measures (Table 2.4). Currently, electricity prices are ratified by the central and provincial governments. According to the existing rules, the electricity price consists of generation price, retail price as well as the transmission and distribution price, though the T&D price is just the difference between the former two and an independent T&D tariff is not in place. Before 2000, the generation price was set on a unit-wise base with a guaranteed return and annual operation hours. To promote competition and reduce administration complexity, the regional benchmarking price for coal power generation was introduced in 2004.

The retail price is also known as the catalogue price, which categorises users with different prices. Generally, commercial users, common industrial users, lighting users (non-household) and big industrial users pay higher electricity rates than the average level, while agricultural users, household users and irrigation users get the subsidised price. In a word, economic growth and social equity are a major concern of the government when setting retail prices, which seriously confuse its economic function. In some provinces, time-of-use tariffs, interruptible load tariffs, etc., which are the key pricing mechanisms for demand-side-management (DSM), are experimentally tested.

To promote the development of renewable energy and energy conservation, renewable energy surcharges and tiered tariffs for households are gradually being implemented in China.

Even in the existing highly planned system, there exists potential for efficiency improvement by substituting the generation of small coal units with more efficient and cleaner ones, if a market for generators to trade their generation rights is in place. The grid companies established a trading platform in 2008 and promoted the generation rights trading among generators, which brings forth a somewhat positive EE effect, though it is difficult to quantify it with the available data.

2.4 Energy Efficiency Achievements in China's Power Sector

In 2009, the power sector alone consumed about 46 % of the total coal supply, emitted 42.8 % of the national SO_2 emissions and 50 % of the CO_2 emissions. With technical innovation, optimisation of the generation mix and improved grid management, the power sector has made progress in EE gains. In 2012, the share of coal power in the capacity mix was down to 71.5 %. The heat rate of coal power generation was down to 325 gce/kWh from 385 gce/kWh in 1998, while the own consumption rate also declined. In this section, the EE achievements are briefly analysed in three perspectives: power generation, grid and final consumption.

2.4.1 Power Generation

Energy conservation in power generation mainly comes from lowering the heat rate and own electricity consumption rate, and improving the generation mix. In 2012, coal power accounted for 71.5 % of total generation capacity, down by 6 % from 2007 (Table 2.5). From 2006–10, a total of 77.25 GW of small coal plants were shut down (Table 2.6), while in the 12th FYP period (2011–15) another 50 GW were expected to close. However, thanks to active EE measures and the efforts of the top state-owned generators (Fig. 2.1), in 2012 the heat rate of the power supply

Table 2.5 Power generation capacities in China, 2007–12

Year	2007	2008	2009	2010	2011	2012
Total installation (GW)	718	793	874	966	1063	1147
Thermal power (GW)	556	603	651	710	768	820
Thermal share (%)	77.40	76.05	74.60	73.40	72.31	71.50
Hydropower (GW)	148	173	196	216	233	249
Wind power (GW)	4	8.4	17.6	29.6	46.2	61.4
Nuclear power (GW)	8.9	8.9	9.1	10.8	12.6	12.6

Source SERC (2008–2013)

Table 2.6 Closure of small coal power generation during 2006–10

Year	2006	2007	2008	2009	2010	Total
Closed capacity (GW)	3.14	14.36	16.68	26.17	16.90	77.25

Source SERC (2008–2011)

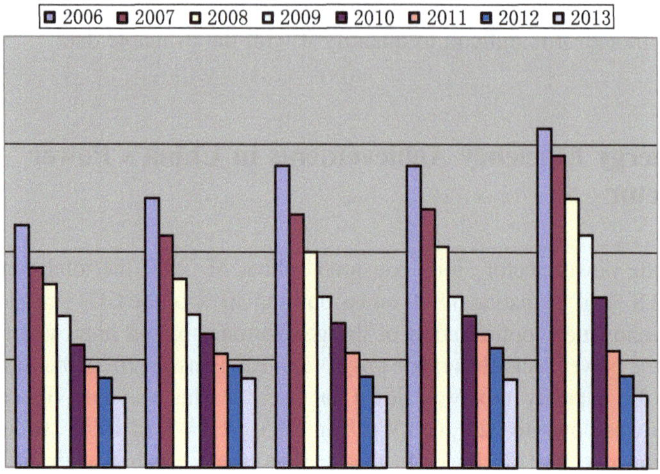

Fig. 2.1 Heat rates of power supply in China's top five state-owned generators. *Source* China Electric Power Yearbook, 2006–13

Table 2.7 Heat rate of power supply in China's power sector

Year	2007	2008	2009	2010	2011	2012	2015	2020	International leading level
Heat rate (gce/kWh)	356	345	340	333	329	325	*323*	*320*	312

Sources SERC (2008–2013); the values for years 2015 and 2020 are planned numbers

Table 2.8 Own consumption rates of coal power plants, 2007–12

Year	2007	2008	2009	2010	2011	2012
Rate (%)	5.83	5.90	5.76	5.43	5.39	5.10

Source SERC (2008–2013)

was down by 8.7 % while the own consumption rate was down by 0.56 % as of 2007 (Tables 2.7 and 2.8).

2.4.2 *Power Grid*

The EE effect in the power grid is manifested in decreased line loss through adjusting the layout of the power grid, reducing active power line loss in T&D, compensating inactive power, smoothing the load curve, and promoting trans-regional power exchange, etc. In 2010 the line loss rate decreased to 6.53 %, which is a little higher than that of the US (6.1 %), but lower than that of the UK (7.7 %), France (7.2 %) and Canada (10.5 %) (Table 2.9).

Table 2.9 Line loss rates in China's power grids, 2007–12

Year	2007	2008	2009	2010	2011	2012
Line loss rate (%)	6.97	6.79	6.72	6.53	6.52	6.42

Source SERC (2008–2013)

2.4.3 Power Utilisation

Green lighting, energy efficient appliances and other EE measures have major energy conservation effects. According to a study of the China National Institute of Standardization (CNIS 2012), the popularisation of 15 typical energy efficient appliances, including motors, air conditioners and others, resulted in electricity savings of 5.8 TWh, or 2.16 Mtce. But compared with industrialised countries, there exists a big gap in the deployment of DSM or energy service in China.

2.5 Energy Efficiency Opportunities in China's Power Sector During 2010–20

In this section, we use the Chinese government's official energy planning as a baseline scenario. Then we compile an ambitious scenario by considering the potential of alternative energy resources and EE improvements. By comparing these two scenarios, the EE opportunities in the sector during 2010–20 are identified.

2.5.1 Baseline Scenario

China's relatively rich coal resources mean that coal will dominate the country's energy supply structure for several more decades. According to the State Council of China (2013), in 2015 the total generation capacity was to reach 1,480 GW, among which 68.65 % was thermal power (1,016 GW). According to the NDRC (2007a–d), NEA (2012b) and CEC (2012), in 2020 the total generation capacity will reach 1,943 GW in China, among which 61.91 % will be thermal power (1,203 GW) (Tables 2.10 and 2.11).

Table 2.10 Baseline power planning scenario in China in 2015 (GW)

Generation type	Hydro	Wind	Nuclear	Solar	Biomass	Gas	Coal
2015	290	100	40	21	13	56	960

Source SCC (2013)

Table 2.11 Installed capacity in related plans and the baseline

Relevant plans	Time issued	Hydro	Wind	Nuclear	Solar	Biomass	Gas	Coal
NDRC (2007d)	2007.8	300	30	–	1.8	30	–	–
NDRC (2007c)	2007.10	–	–	40	–	–	–	–
CEC (2012)	2012.3	330	180	80	25	5	43	1160
NEA (2012b)	2012.8	420	200	–	50	–	–	–
Baseline	–	420	200	40	50	30	43	1160

Source NRDC (2007c, d), CEC (2012), NEA (2012b) power planning scenario in China in 2020 (GW)

According to the baseline scenario, in 2015 the share of coal power in the total generation mix will be lowered to 68.86 %, and in 2020 it will be further reduced to 59.70 %. Compared with the 2010 level, the substitution of coal power by clean and renewable energy will result in energy conservation of 394 Mtce by 2015. Relative to the 2015 level, the substitution of coal power will result in additional energy conservation of 338 Mtce by 2020 (Tables 2.12 and 2.13).

Table 2.12 Pollutant emission factors of coal power generation

Pollutant	CO_2	SO_2	NO_x
Emission factor (g/kwh)	900	2.71	2.68

Sources CO_2 emission factor is sourced from IEA (2011); SO_2 and NO_X factors are sourced from NRDC (2015). Active end-of-pipe emissions control (including retrofitting) has effectively lowered the emission factors of SO_2 and NO_x form coal power generation in China since 2002. Because the time horizon of our study is 2010–20, we use 2010 factors for estimation

Table 2.13 Energy saving and emissions abatement resulting from reducing the coal power share in China, 2010–20

Year	Total installation (GW)	Coal capacity (GW)	Coal share (%)	Saved primary energy (Mtce)	CO_2 abatement (Mt)	SO_2 abatement (Mt)	NO_x abatement (Mt)
2010	966	710	73.40				
2015/10	1480	960	64.86	394	1125	3.39	3.35
2020/15	1943	1160	59.70	338	990	2.98	2.95

Source Authors' calculation. The annual operation of coal-fired power plants is set as 5,500 h

2.5.2 *Energy Efficiency Scenario*

2.5.2.1 Contribution from Technical and Operational Improvements

Though the operation efficiency of China's coal power plants and power grids is comparable with advanced global standards, there still exists a substantial gap in terms of best practices. Assuming that by 2020 the own consumption rate reaches the Japanese level in 2002 (3.45 %), as does the line loss (4.75 %) (SERC 2010; SGCC 2010), we estimate the EE contribution from technical and operational improvements (Table 2.14). It is evident that the combined effects of reducing the own consumption rate and line loss rate can bring forth primary energy savings of 50.49 Mtce and avoidance of 148.98 Mt of CO_2 emissions, compared to 2010 levels.

2.5.2.2 Contribution from Improvements in the Generation Mix

Structural improvement of coal power. Efficiency improvements and structural changes in coal power can contribute to low-carbon development of the power sector. Based on the statistics of the CEC (2013a), the structure of coal power plants

Table 2.14 The potential of energy conservation and emissions abatement from technical and operational improvements

Contribution	Primary energy conservation (Mtce)	CO_2 abatement (Mt)	SO_2 abatement (Mt)	NO_x abatement (Mt)
Auxiliary rate (3.45 %)	25.09	148.98	0.45	0.44
Line loss (4.75 %)	25.40			
Total	50.49			

Source Authors' calculation

Table 2.15 Statistics on coal power plants in China, 2010

Capacity type (MW)	Subtotal (GW)	Share (%)	Annual operation (h)	Heat rate (gce/KWh)
Unit ≥ 1000	58	9.31	5400	292
600 ≤ unit < 1000	247	39.65	5122	313
300 ≤ unit < 600	239	38.36	4525	322
200 ≤ unit < 300	42	6.74	4451	342
Unit < 200	37	5.94	4713	365

Source CEC (2013b)

and their major technical economy indices are provided in Table 2.15. The units above 300 MW account for 76 % of all the coal power plants, but small units (below 200 MW) still account for about 6 %.

Currently, SC is the dominant technology in the coal power generation units. Compared with SC, USC performs better with 2–3 % more in generation efficiency and its heat rate is less than 290 gce/kWh. Because of its superiority in generation efficiency and environmental performance, USC will become the mainstream option in the future. Supposing that all new installations of coal units are 600 MW SC or above, and that most of the old units below 300 MW are replaced by 600 MW SC units, the potential energy conservation and emissions abatement by structural adjustment in coal power are estimated in Table 2.16.

According to our estimate, with radical substitution, the share of units above 300 MW could be increased to 98 % in 2020. The total potential energy conservation in coal power by structural adjustment and retrofitting could reach 33.6 Mtce and result in more than 99.14 Mt of abatement in CO_2 emissions.

Ambitious clean generation. Extra EE improvements can be realised with ambitious clean energy development. The role of nuclear power in clean energy development has been highlighted by many countries. In all the top-10 energy consumption countries, the contribution of nuclear power is well above 15 %. For instance, in 2005 the share of nuclear in total power generation was 77.6 % in France, 28.1 % in Germany, 25 % in Japan, 23.7 % in the UK, 20 % in the US and 16.5 % for Russia (NDRC 2007d). As the largest energy consumer in the world, China lags far behind in nuclear power development. China has an excellent safety record in the operation of nuclear power (zero Grade 2 incidents until the end of 2013) and has abundant siting resources for nuclear power; while the proactive safety system of the third-generation nuclear technology and the reuse of spent nuclear fuel could make nuclear power even more successful. Actually, the installation of nuclear power could be increased to 65–5 GW over that in the BAU planning.

Though China has become the world's largest wind power developer in terms of installed capacity, its wind capacity is negligible relative to the abundant resource endowment. Thanks to remarkable learning-by-doing, wind power has become economically competitive with coal power in Southern China where the price of

Table 2.16 Potential of energy conservation and emissions abatement by structural adjustment and retrofitting of coal power, 2020

Capacity type (MW)	Share (%)	Subtotal (GW)	Heat rate (gce/kWh)	Primary energy conservation (Mtce)	CO_2 abatement (Mt)	SO_2 abatement (Mt)	NO_x abatement (Mt)
Unit ≥ 1000	9.31	108.0	292(28)	16.33	99.14	0.30	0.29
600 ≤ unit < 1000	51.21	594.09	313(7)	21.30			
300 ≤ unit < 600	38.36	444.98	322(−2)	−4.03			
Total	98.88	1147.07	–	33.60			

Source Authors' calculation. Numbers in the brackets stand for improvement in heat rate by retrofitting

Table 2.17 Potential of alternative power generation in the cleaner scenario in 2020

Alternative power	BAU scenario share (%)	Cleaner scenario share (%)
Nuclear	2.06	3.35
Wind	10.29	12.87
Solar	2.57	5.15
Total	14.92	21.37
Reduced coal power share (%)		6.45
Primary energy conservation (Mtce)		131.68
CO_2 abatement (Mt)		370.36
SO_2 abatement (Mt)		1.12
NO_x abatement (Mt)		1.10

Source Authors' calculation

coal power generation is high. Our estimate is that with further learning-by-doing, wind power will become fully competitive in two to three years. Therefore, in the cleaner scenario, wind power is expected to experience rapid growth and reach 250 GW in 2020. Solar power holds a negligible position in the BAU scenario with 50 GW of capacity installations in 2020. However, if China can implement strong facilitating policies, solar power will very likely take off. Accordingly, in the cleaner scenario, it is expected that solar capacity will reach 100 GW in 2020 (Table 2.17).

2.5.2.3 Contribution from DSM

DSM was introduced in China in the 1990s. But its actual effect is minimal in the existing deployment pattern of command-and-control. However, DSM is widely deployed in developed countries with great success due to the market mechanism. In California for example: with 30 years of effort, DSM has successfully decreased the peak load in the state by 12 GW, approximating 15 % of the power load (Baskette et al. 2006). During 2000–01, DSM effectively reduced the power demand of California by 6 %. There have also been similar success in the UK, France, Japan and Denmark. If China can utilise the market mechanism to implement DSM as these countries have done, we estimate that DSM can save at least 3 % of electricity demand (or 260 TWh) in 2020 (NRDC 2014). The contribution is estimated in Table 2.18.

2.5.2.4 Total Potential of Energy-Efficient Scenarios

According to the above analysis, 295.85 Mtce of primary energy could be saved with these EE measures in China's power sector by 2020, and a total of 852.48 Mt CO_2 emissions could be avoided (Table 2.19). According to the study by

Table 2.18 Potential of energy conservation and emissions abatement by DSM in 2020

	Electricity supply (TWh)	Avoided demand (TWh)	Primary energy conservation (Mtce)	CO_2 emissions (Mt)	SO_2 emissions (Mt)	NO_x emissions (Mt)
Without DSM	8,720	260	80.08	234	0.70	0.70
With DSM	8,460					

Source Authors' calculation

Table 2.19 Total potential of energy conservation and emissions abatement in the EE scenario

Contribution	Primary energy conservation (Mtce)	CO_2 abatement (Mt)	SO_2 abatement (Mt)	NO_x abatement (Mt)
Operation improvement	50.5	148.98	0.45	0.44
Coal power	33.60	99.14	0.30	0.29
Clean energy	131.68	370.36	1.12	1.10
DSM	80.08	234.00	0.70	0.70
Total	295.85	852.48	2.57	2.53

Source Authors' calculation

Yuan et al. (2014a, b), in 2020 total emissions would reach around 9,000 Mt if China achieves its GDP CO_2 intensity reduction target of 45 % on the baseline of 2005 levels. By taking active measures, the power sector alone can contribute a significant portion to the CO_2 abatement target in China.

2.6 Challenges and Policy Implications

Though huge EE potentials have been detected in China's power sector, there are big challenges to fully deploy them. In the future, the EE potential will come more from the power grid and demand sides, instead of from power generation (especially coal power) as in the past. Closure of small coal units and substitution with large SC or USC units contributed to the largest share of EE gains in the past. In the coming years, although the potential of structural adjustments in coal power is still substantial, their deployment will be very challenging for the central government, given the past strong opposition from the generators and local governments when closing down small coal units. A strong implication is that focus of EE measures will shift from technical efficiency to economic efficiency. Furthermore, a transition in the deployment mechanism from command-and-control (CAC) that the Chinese government is used to, to market-based economic measure is necessary.

The first policy implication is that China can no longer ignore the externality of coal and must gradually internalise it into the cost of power generation (NRDC 2014). In this way, the inefficient coal power generators, who must incur higher costs by emissions taxes and/or carbon taxes, will find themselves in a disadvantageous position compared with efficient and clean generators.

The second implication is that the Chinese government must loosen its tight grip on price control and let the market provide proper price signals. A two-part pricing mechanism for power generation—with capacity pricing to cover the huge long-run capital investment and power pricing to cover the short-run variable cost (i.e. marginal cost pricing)—can be employed. With such a generation pricing mechanism in place, more efficient coal generators and renewable generators will find themselves in an advantageous position in competition because of their lower marginal cost. In this way, China can put an end to the existing feed-in-tariff policy for wind and solar power, and switch to unified market rules and investment subsidies for renewable generators. Meanwhile, genuine market competition requires that the government lift its ban on private and foreign investors, and establish a level playing field for all in the sector.

The third implication is that the government must regard energy efficiency as a precious resource in power planning, and establish a strong market for deploying energy services. The fourth and closely related implication is that the industrial structure of the power sector must be radically changed. Though grid operators are obliged with energy efficiency responsibilities, they have vested interests in selling more electricity. Thus, total separation of retail businesses from transmission and distribution is complicated and will take long time. A good starting point would be to define the grid as a pure public utility and then redefine its incentive structure through tariff reforms and stronger regulations.

2.7 Conclusions

In a highly planned system, China's power sector achieved some results in energy efficiency through direct government regulations to shut down small coal units and by radical progress in coal power technologies. In this chapter, the analysis showed that the energy efficiency potential will shift from coal power to renewable energy, the power grids and the demand side. In turn, a transition from command-and-control incentives to market-based economic incentives was identified as crucial. To fully explore the energy efficiency potential, the following policies are proposed: imposing emissions taxes or fossil energy taxes to offset the negative externality of coal power; conducting market reforms, including price reform in particular; formulating legislation for integrated resource planning; and finally, restructuring the power sector and redefining the grid company as a pure public utility.

References

Baskette, C., B. Horii et al. (2006) "Avoided Cost Estimation and Post-Reform Funding Allocation for California's Energy Efficiency Programs". *Energy* 1: 1084–99.

CEC (China Electricity Council) [中国电力联合会] (2012) *Power Sector Research Report for the 12th FYP Period* [中国电力联合会,2012.电力行业十二五规划研究报告]. http://www.cec.org.cn/zhuanti/2011nianzhuantihuigu/dianlixingyeshierwuguihua/.

CEC (China Electricity Council) (2013a) *EE Data Tables for 600 MW Coal Power Generation Unit in 2012* [2012年度全国火电600 MW级机组能效对标基础信息表]. http://kjfw.cec.org.cn/kejifuwu/2013-04-07/99877.html.

CEC (China Electricity Council) (2013b) *Statistical Data for Power Sector in 2012* [2012年电力统计基本数据]. http://www.cec.org.cn/guihuayutongji/tongjxinxi/.

CNIC [中国标准化研究院] (2012) *White Paper for the Energy Efficiency Status of China Energy-Use Products* [中国用能产品能效状况白皮书]. Beijing: China Zhijian Publishing House [中国标准出版社].

International Energy Agency (IEA) (2011) *CO_2 Emissions from Fuel Combustion*. Paris: IEA.

NDRC [发改委] (2007a) *Rules on Energy Conservation Dispatch (Trial)* [节能发电调度办法 (试行)]. http://www.sdpc.gov.cn/zcfb/zcfbqt/200708/t20070828_156042.html.

NDRC (2007b) *Rules on the Closure of Small Coal Power Generation Units* [关于加快关停小火电机组的若干意见]. http://www.sdpc.gov.cn/zcfb/zcfbqt/200701/t20070131_115037.html.

NDRC (2007c) *Notice on the Medium-and-Long-Term Plan for Nuclear Energy Development* [核电中长期发展规划]. http://www.ccchina.gov.cn/WebSite/CCChina/UpFile/2007/2007112145723883.pdf.

NDRC (2007d) *Notice on the Medium-and-Long-Term Plan for Renewable Energy Development* [可再生能源中长期发展规划]. http://www.ccchina.gov.cn/WebSite/CCChina/UpFile/2007/20079583745145.pdf.

NDRC (2008a) *Notice on Grid-Access Price for Wind Power* [关于风电上网电价通知].

NDRC (2008b) *Policy on Forbidding the Preferential Price in Energy-Intensive Industries* [关于高耗能行业电价的通知].

NDRC (2009) *Notice on Improving Grid-Access Price for Wind Power* [关于提高风电上网电价通知].

NDRC (2010) *Pollutant Emissions Factors of Coal Power Generation* [火电行业CO2排放因子推荐数据表].

NDRC (2011a) *Notice on Price for Household Electricity Consumption* [关于居民生活用电实行阶梯电价的指导意见]. http://www.sdpc.gov.cn/fzgggz/jggl/zcfg/201111/t20111130_448436.html.

NDRC (2011b) *Rules on the Development of Combined Heat and Power Generation* [关于发展热电联产的若干规定]. http://xmecc.xmsme.gov.cn/2011-8/2011812161937.htm.

NDRC (2012) *Policy on the Management of Renewable Energy Fund from Price Surcharge* [可再生能源电价补贴和配额交易方案]. http://www.sdpc.gov.cn/fzgggz/jggl/zcfg/201212/t20121204_517159.html.

NEA [国家能源局] (2005) *Notice on the Renewable Energy Law* [可再生能源法]. http://www.gov.cn/ziliao/flfg/2005-06/21/content_8275.htm.

NEA (2007) *Notice on Energy Conservation Law* [节约能源法]. http://www.gov.cn/ziliao/flfg/2007-10/28/content_788493.htm.

NEA (2012a) *Evaluation of Energy Consumption Status in Power Plants* [电厂能效评价]. http://www.nea.gov.cn/2012-02/10/c_131402684.htm.

NEA (2012b) *Renewable Energy Development Plan for the 12th* FYP [可再生能源发展"十二五"规划]. http://www.nea.gov.cn/2012-08/08/c_131767651.htm.

NEA (2014) *Guiding Instruction of 2014 Energy Work* [2014年能源工作指导意见]. http://www.cec.org.cn/yaowenkuaidi/2014-01-26/116202.html.

NRDC (Natural Resource Defense Council) [自然资源保护委员会] (2014) *The True Cost of Coal, 2012* [煤炭真实成本 (2012) 研究报告].

NRDC (Natural Resource Defense Council) (forthcoming) *Coal Consumption Cap and Implementation Policy in the Power Sector* [电力部门煤炭消费总量控制与政策研究报告(即将发布)].

NSBC (National Statistical Bureau of China) [国家统计局] (2013) *China Statistical Book 2013* [2013.中国统计年鉴]. http://www.stats.gov.cn/tjsj/ndsj/2013/indexch.htm.

SCC (State Council of China) [国务院] (2013) *Energy Development Plan for the 12th FYP* [能源发展"十二五"规划]. http://www.gov.cn/zwgk/2013-01/23/content_2318554.htm.

SERC (State Electricity Regulatory Commission) [国家电监会] (2008) *Notice on Energy Conservation and Emissions Abatement for 2007* [2007电力行业节能减排情况通报]. http://wenku.baidu.com/link?url=Su70bqh6eazqN7F3UE4cXaOuuRplKbab0IH7Z5Q64SAMKpuBzOv0DUIGaMp_S47yotw0fmfhqWKcNV31zd9IiyZuAtz3xa9aRTQH91rLwH_.

SERC (State Electricity Regulatory Commission) (2009) *Notice on Energy Conservation and Emissions Abatement for 2008* [2008电力行业节能减排情况通报]. http://www.zhb.gov.cn/gkml/hbb/gwy/200911/W020091102480145922089.doc.

SERC (State Electricity Regulatory Commission) (2010) *Notice on Energy Conservation and Emissions Abatement for 2009* [2009电力行业节能减排情况通报]. http://www.gov.cn/gzdt/2010-11/01/content_1735597.htm.

SERC (State Electricity Regulatory Commission) (2011) *Notice on Energy Conservation and Emissions Abatement for 2010* [2010年及"十一五"电力行业节能减排情况通报]. http://energy.people.com.cn/mediafile/201110/26/P201110261109182300311272.doc.

SERC (State Electricity Regulatory Commission) (2013) *Notice on Energy Conservation and Emissions Abatement for* 2012 [2012年电力行业节能减排情况]. http://www.cec.org.cn/yaowenkuaidi/2013-02-06/97216.html.

SGCC (State Grid Corporation of China) [国家电网公司] (2010) *International Comparison of Grid Line Loss* [电网线损的国际比较].

Yuan, J., J. Shen, L. Pan et al. (2014a) "Smart Grids in China". *Renewable and Sustainable Energy Reviews* 37: 896–906.

Yuan, J., Y. Xu, X. Zhang et al. (2014b) "China's 2020 Clean Energy Target: Consistency, Pathways and Policy Implications". *Energy Policy* 65: 692–700.

Chapter 3
Efficiency in China's Power Sector: Evidence from a Large Dataset of Power Plants

Chunbo Ma

Abstract China is making a green refurbishment of its economy. Keystones in China's greening process include: a structural shift to high-end manufacturing, services and research and development, a transition to renewable energy technologies, and a large-scale retrofitting of its conventional energy sector. Perhaps nowhere is this conservation and transition effort greater than in the electricity sector. China has introduced a number of command and control, as well as market-based policy instruments, to clean up its conventional energy supply. This chapter reviews and analyses the effectiveness of some of the policies introduced in recent years, drawing on recent empirical literature using plant-level data. It also identifies a number of challenges and opportunities in the sector for the foreseeable future.

Keywords Efficiency · Deregulation · Electricity · China · Data envelopment analysis · Stochastic frontier analysis

3.1 Introduction

China's growing demand for energy seems insatiable; however, the dismal air quality across much of the country is a constant reminder of its heavy reliance on coal (and to a lesser extent on other fossil fuels). It burns more coal than the rest of the world combined and emits more carbon and sulphur dioxide than any other country (BP 2013; Boden et al. 2010; Smith et al. 2011). Energy consumption and its consequences for health and the environment are high on the Chinese political agenda (Hughes 2012). About half of the fossil fuels consumption as well as the associated pollution come from fossil fuel-fired power generation. In fact, much of

Dr. Chunbo Ma, Assistant Professor, School of Agricultural and Resource Economics, Centre for Environmental Economics and Policy, University of Western Australia; Email: chunbo.ma@uwa.edu.au.

B. Su and E. Thomson (eds.), *China's Energy Efficiency and Conservation*,
SpringerBriefs in Environment, Security, Development and Peace 30,
DOI 10.1007/978-981-10-0737-8_3

recent and future policy attention has been paid to the power sector to address the dismal air quality. For instance, China has recently announced bans on new coal-fired power generation in Beijing and its two largest industrial areas—the Yangtze Delta region near Shanghai and the Pearl River Delta region of Guangdong province. Governments at all levels are also taking measures to further improve the efficiency of existing power plants. Given the sheer size of China's power generation, how this sector fares in future will also largely determine our planet's climate.

China's energy planners have realised that improving energy efficiency may be the easiest way to promote economic growth while controlling pollution. The government has effectively made energy efficiency one of its major climate policies, and has been making great efforts to restructure its electricity sector while refurbishing its conventional energy supply (Aldhous 2005; Tollefson 2008). The Chinese government has introduced a number of market-restructuring reforms in its electricity sector, including: the *Investment Decentralization* in 1986; the first unbundling reform (*Unbundling of Government Admin and Business Operations*) in 1997; and the second unbundling reform (*Unbundling of Generation and Transmission*) in 2002 (Ma/He 2008). Most existing studies on China's market restructuring reforms have examined the impact of the 2002 unbundling reform on power plants' efficiency. However, such assessment is no easy task as the sector has been adapting to other regulatory changes over roughly the same period. The government has made great efforts to retrofit its generation fleet by introducing a *Small-Unit Shutdown Mandate* (*SUSM*) in 1999, a *Promoting the Big and Quashing the Small Policy* (*PBQSP*) in 2007, as well as environmental mandates such as the *Flue Gas Desulfurization Mandate* (*FGDM*) in 2006.

The primary aim of this chapter is to review what we have learned about the impacts of these changing policy parameters, with a focus on the relative contributions of various policies which could have very different implications for future efficiency trajectories. Such understanding is pivotal for informed scenario calibration of future national and international energy consumption and greenhouse gas (GHG) emissions, and climate policymaking. This chapter provides a unique analysis of these changing policy parameters based on a large panel set of power plants. The next section presents the observed efficiency trajectory over the 1997–2010 period and associated technical characteristics. Section 3.3 then reviews the impact of the unbundling reform on the efficiency of China's power plant sector drawing from the existing literature. Section 3.4 examines a number of retrofitting mandates and potential impacts on power plants' efficiency. Section 3.5 provides an integrated analysis of these changing policy parameters. The last section concludes with a discussion of opportunities and challenges.

3.2 Newer, Larger, Cleaner and More Efficient Generation

An oft-cited statistic is that the Chinese bring a new power plant on line every week or two; less appreciated is the fact that the new plants consist of larger generation units and employ state-of-the-art combustion technology and emission control technology while older and smaller generation units are shut down. The last one and a half decades have witnessed substantial efficiency gains in China's power sector. Figure 3.1 shows the plant-level heat rate during 1997–2010 with a trend line, where we can identify three significant patterns: (1) the heat rate has improved on average, indicating a shift of the efficiency frontier; (2) the plant-level heat rate has converged over time, which is evidence for the laggards catching up with the leaders; and (3) the generation units are newer, larger and cleaner, which may also contribute to efficiency variations. Empirically, researchers have studied the operational efficiency rather than the heat rate as the former reflects the multi-factor productivity. Such multi-factor productivity analysis also confirms patterns of improving as well as converging efficiency. Figure 3.2 presents a kernel distribution of technical and scale efficiency (TSE) which has been estimated as a multi-factor efficiency indicator. The distribution has not only shifted to the right over time, indicating an improving trend, but has also become less dispersed, reflecting a

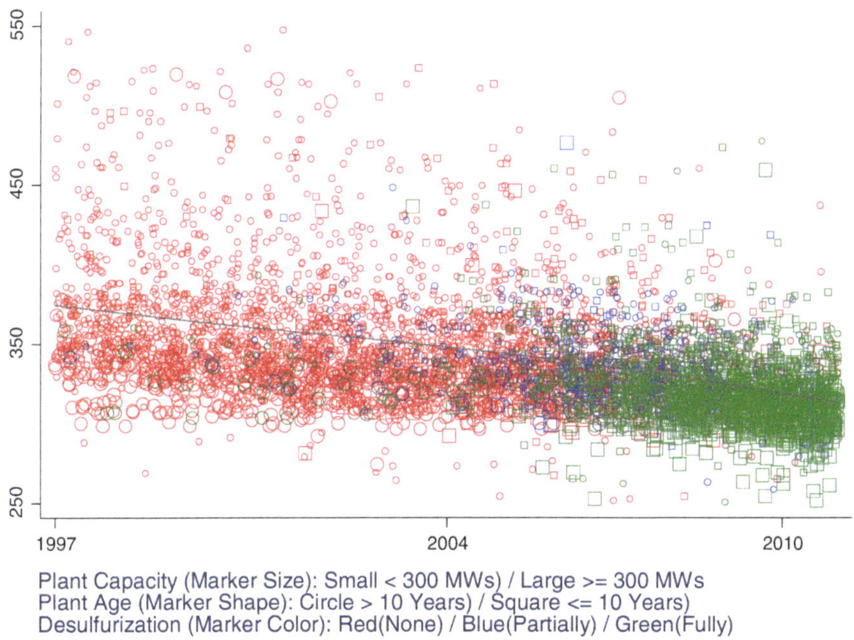

Fig. 3.1 Plant heat rate (grams of standard coal equivalent per kilowatt hour). *Source* Ma/Zhao (2015)

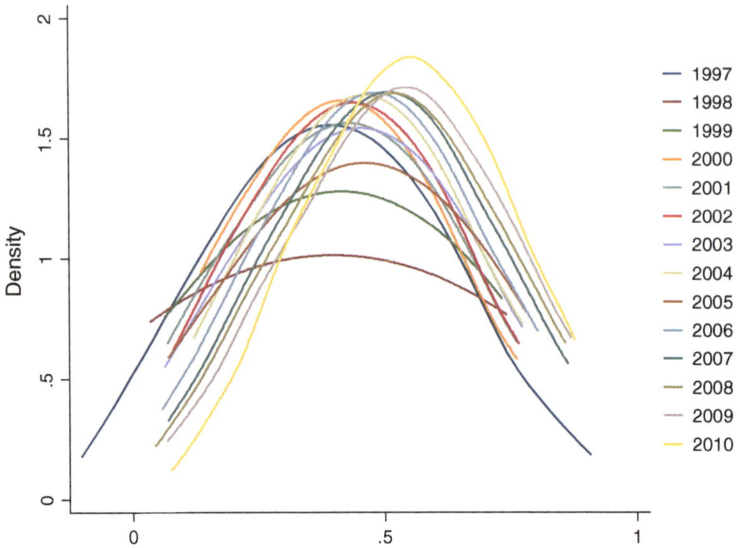

Fig. 3.2 Kernel distribution of plant-level TSEs 1997–2010. *Source* Ma (2012)

converging process. These observed patterns may be attributed to a number of regulatory changes including the market-restructuring reform and technology mandates, among others. Some efficiency gain may be a result of dynamic incentives while others may well just be one-off. It is thus of relevance to gauge the relative contributions of these changing policies if modellers and policymakers want to calibrate projections based on these observations of historical trajectories.

3.3 Market-Restructuring Reforms

Many electricity markets around the world have been decentralised and restructured with the aim of increasing competition and improving operational efficiency. A common theme of China's three market-restructuring reforms in 1986, 1997 and 2002 was to improve the operational efficiency of the sector. However, the first two reforms brought only weak incentives for improvement as most power plants were still dominated with state shares and faced only a soft budget constraint. Furthermore, the entire electric utility was still vertically integrated with generation, transmission and distribution controlled by the State Power Corporation (SPC). Professor Yuan provides a more detailed discussion of these reforms in a different chapter.

Most existing studies focus on the unbundling reform in 2002 as it was intended to fundamentally change the market structure and introduce genuine competition. The reform aimed to remove the monopoly of the SPC and introduce competition on the generation side. The vertically integrated SPC was dismantled into 11 new corporations including five generation corporation groups, two grid corporations and four auxiliary corporations. Each of the five generation groups manages a large number of power plants. It was also hoped that the right to dispatch power would be based on economic efficiency and merit order rather than political factors such as protection of state-owned assets and employment. This is to be achieved by establishing a competitive wholesale electricity market. The competitive wholesale market is generally viewed as the critical part of the reform as it will create a dynamic incentive for all plants to improve efficiency. However, a fully functioning regional or national wholesale market is yet to be established. Researchers have raised questions about whether and to what extent the electricity sector has benefited from this rather incomplete market restructuring.

Du et al. (2009) were the first to investigate the impact of the unbundling reform in 2002 on China's electricity generation efficiency using two cross-sectional firm-level national survey data collected in 1997 and 2004; they found significant input efficiency improvement in non-fuel materials but not in fuel input. This was further confirmed in Du et al. (2013). Yang/Pollitt (2009) also examined the productivity performance of Chinese coal-fired power generation based on a single cross-sectional sample of 221 power plants in 2002; however, the focus of the paper was primarily on the relative performance of different nonparametric data envelopment analysis (DEA) models rather than the impact of regulatory reforms. Gao/Biesebroeck (2014) provided the first parametric firm-level panel evidence and confirmed the efficiency improvement due to the 2002 unbundling reform using a sample from 1998 to 2007. More recently, Zhao/Ma (2013) conducted a semi-parametric analysis on the productivity of China's largest coal-fired power plants which also confirmed the efficiency gain due to the unbundling reform.

A consensus has emerged that even though the reform is incomplete, there is still significant efficiency gain due to increased competition. Following the unbundling of generation from the grid, reducing fuel costs and improving operational performance have increasingly become the focal subject of management practice in recent years. The initial design of the reform sent a strong signal for marketisation, which has already been playing a role in guiding the management practice and investment choices of power companies. This is similarly addressed in Zhao/Ma (2013), and Gao/Biesebroeck (2014).

In contrast to the strong evidence for the unbundling reform, the existing literature has been mostly silent about the contributions of other related policy measures, including particularly the retrofitting mandates implemented during the same period. All the aforementioned studies employed a Difference in Difference (DID) approach to assess the impact of the 2002 unbundling reform; however, the validity of DID hinges crucially on sufficient control for entangling policies. Given that the retrofitting process may have significant impacts on the operational efficiency of power plants and the sector has experienced drastic retrofitting processes in recent years, it is of significance to gauge the relative contributions of different policy measures. This insight will not only help to interpret the significant efficiency gain observed in recent years (Figs. 3.1 and 3.2), but also facilitate improved calibration of future fuel demand and emission scenarios.

3.4 Technology Mandates

3.4.1 Small-Unit Shutdown Mandate (SUSM) and Promoting the Big and Quashing the Small Policy (PBQSP)

A substantial proportion of China's thermal-fired capacity has been provided by small-scale coal- and oil-fired units, where small scale is defined as any unit with a capacity of less than 100 MWs.[1] These small units are generally inefficient in their use of energy and also highly polluting (Cao et al. 2009).[2] The government targeted 50 GWs of small-scale capacity for closure by 2010: the regulatory authority published an annual list of small-scale units to be decommissioned in each year, which was known as the *SUSM*. Implementing the *SUSM* would require that new capacity (equivalent or even larger) be built for replacement. The *PBQSP* aims to encourage power plants to fulfil the replacement with larger and more advanced units believed to be more efficient and less polluting (Fig. 3.3).

As a result of these technology mandates, the technical profiles of China's power plants have changed substantially over the studied period. Figure 3.4 shows the mean utilisation, mean capacity-weighted vintage and mean capacity-weighted unit size of all plants that have a dominant SPC ownership before 2003 and all independent power producers (IPPs). While the average vintage of both kinds of plants was decreasing, the units of the SPC plants are on average five years older than those of the IPPs. This is consistent with the fact that the IPPs have only appeared after the 1986 reform which removed the restrictions on investment and diversified investment channels. There was also a sharp difference in the change of unit size between the two groups of plants. While both seemed to have larger units over time, the growth in unit size was much faster for the IPPs.

It is generally expected that performance eventually deteriorates as a unit ages, and that units of an older vintage are generally less efficient. However, units may go through a break-in period early in their lives, which is usually characterised by a high level of forced outages and derating or cycling of the facility. This means that observed performance may actually improve during these earlier years of operation, indicating a non-linear trend of performance (Joskow/Schmalensee 1987; Pollitt 1995). Figure 3.5 presents scatter plots of TSE versus capacity-weighted vintage with a lowess smoother for each sampled power plant where vintage is defined as

---

[1]According to the new program, the following categories of thermal units are targeted for closure: (1) units below 50 MW; (2) units below 100 MW that have been operating for over 20 years; (3) units below 200 MW that have reached the end of their design lives; and (4) units with coal consumption 10 % higher than the provincial average or 15 % higher than the national average.

[2]These small units are generally inefficient and also highly polluting. The average total cost per kilowatt hour for small plants is almost three times the cost for large plants. Most of these units were state-owned and built to serve localities that had in the past experienced severe electricity shortages (Cao et al. 2009).

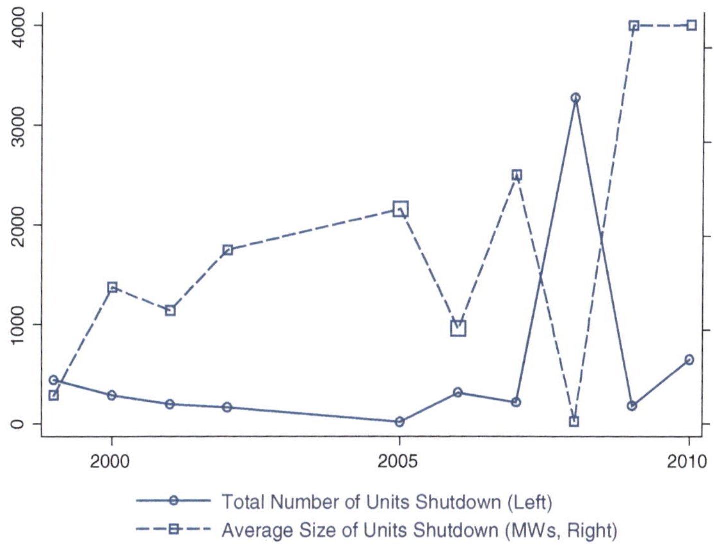

Fig. 3.3 Small thermal units shutdown 1999–2010. *Source* Zhao/Ma (2013), Ma/Zhao (2015)

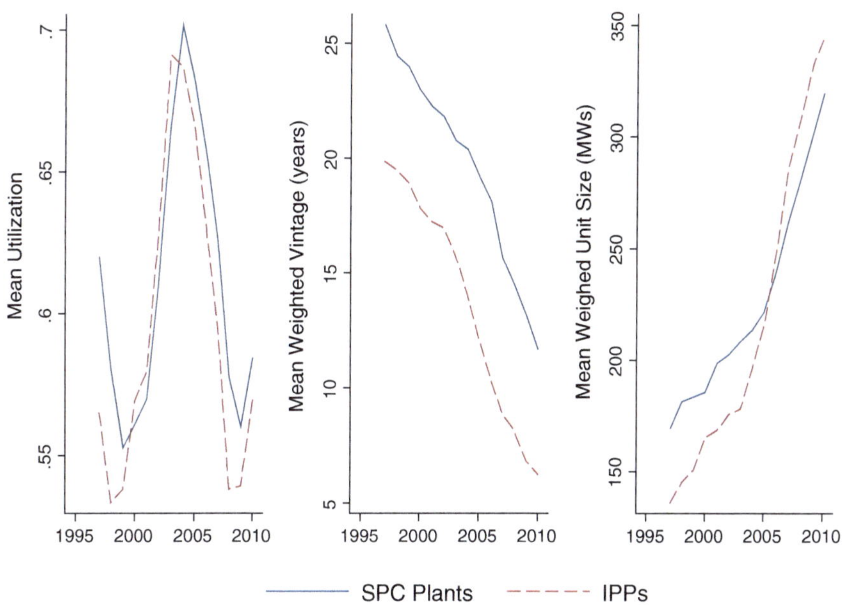

Fig. 3.4 Mean utilisation, capacity-weighted vintage and capacity-weighted unit size. *Source* Ma/Zhao (2015)

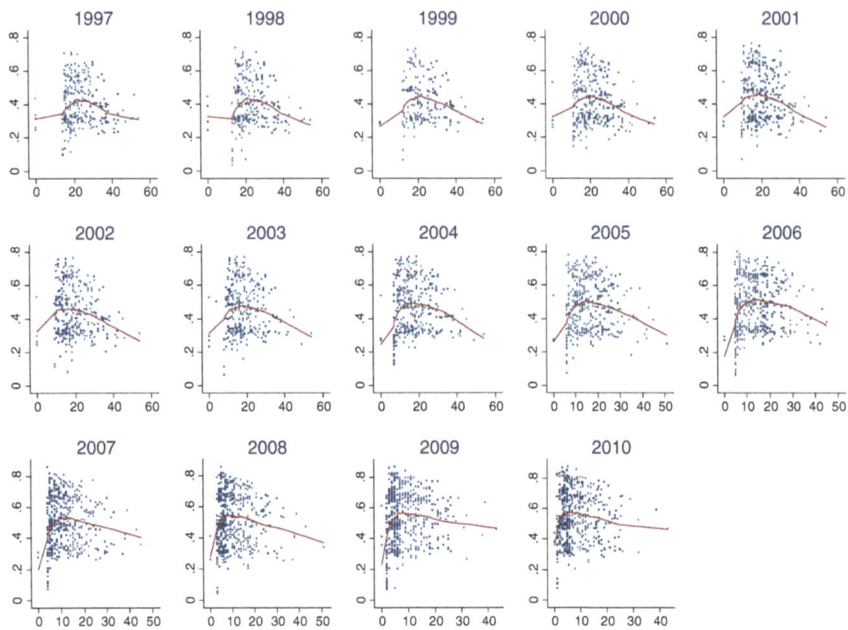

Fig. 3.5 Plant-level TSE versus capacity-weighted vintage. *Vertical axis* represents TSE scores and *horizontal axis* indicates capacity-weighted vintage years. *Source* Ma (2012)

the last calendar year of the sample period (2010) minus the year of initial operation. The plots seem to confirm such a non-linear relationship. However, this non-linearity could well be an artefact of sparse data at the low end (i.e. completely new built power plants). Nevertheless, it is fair to say that older vintage is associated with lower efficiency.

Other things being equal (e.g. steam temperature, pressure and fuel characteristics), a larger boiler should reduce the unit's heat rate; however the advantage of a larger size should be more significant at a small scale than a large scale. This is because larger units have poorer availability than smaller units and the advantage of larger units for heat rate may disappear when the costs of poor availability are factored in (Joskow/Schmalensee 1987). Nevertheless, considering that the typical size of many China's generating units is relatively small, a positive relation between unit size and efficiency is expected. Note that what matters here is the size of the units, not the overall size of the plant. A plant may have multiple units with different size profiles. A plant's unit size can be defined as the average size of all units of the plant weighted by each unit's nameplate capacity. Figure 3.6 presents scatter plots of TSE versus capacity-weighted unit size with a lowest smoother for each sampled power plant, which confirms that the retrofitting process has indeed contributed to observed improvement in operational efficiency.

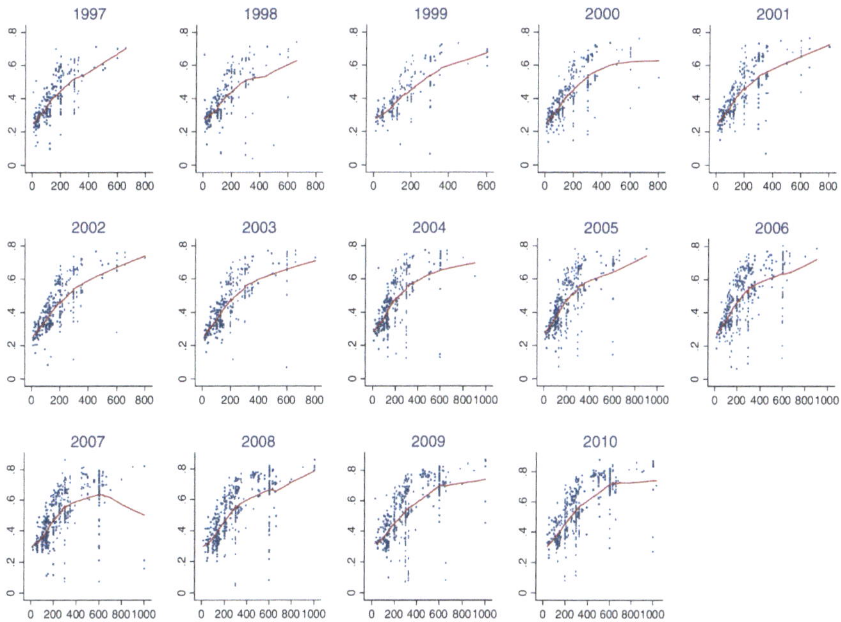

Fig. 3.6 Plant-level TSE versus capacity-weighted unit size. *Vertical axis* represents TSE scores and *horizontal axis* indicates capacity-weighted vintage years. *Source* Ma (2012)

As Zhao/Ma (2013) point out, frequent retrofitting and upgrading may create bias in statistics and efficiency fluctuations in the short term. These biases and fluctuations may be statistically severe if a large number of plants commission new units or decommission old units in a short period of time, which is exactly the case for China's electricity sector in the past few years. If a small-scale unit is decommissioned sometime in the mid-year, the capacity of this unit would most likely not be included in the annual statistic of the plant's total capacity although the decommissioned unit is actually commissioned for part of the year. However, the electricity generated would still be included in the annual statistic of the plant's total generation as it is a flow statistic. This accounting practice would under-report the generation capacity that is actually at work and thus inevitably bias the efficiency score upwards. Similarly, commissioning a new unit in the mid-year would over-report the generation capacity that is actually at work and thus bias the efficiency estimate downwards. However, decommissioning old units or commissioning new units may also drive down the overall efficiency of other units of the same plant as normal operation may be affected and resources may be re-allocated. Empirical analyses of power plant efficiency would have to account for such potential biases and short-term fluctuations.

3.4.2 Flue Gas Desulfurization Mandate (FGDM)

With mounting domestic and international pressure to mitigate GHGs emission, and an urgent need to improve ambient air quality, China is also taking measures to reduce emissions from the electricity sector which include the Preferential Tariff for Desulphurised Generation (2004) and the Flue Gas Desulphurisation Mandate (FGDM) (2006). China now has more flue gas desulphurisation (FGD) capacity than the rest of the world put together (Tollefson 2008). Nitrogen oxides are the next on the clean-up list with the recently introduced Preferential Tariff for Denitrated Generation in 2012. By 2005, FGD equipment had only been installed on 12 % (46.2 GWs) of China's total generation capacity (Cao et al. 2009). However, this was much improved in the past few years due to the FGDM. To meet the sulphur reduction target of the 11th Five-Year Plan, the government scheduled FGD facilities to be installed on a total of 167 GW existing thermal generation units by 2010. In addition, all newly commissioned generation units are mandated to install FGD facilities. Figure 3.7 shows accumulated desulphurisation capacity since 1997.

Environmental conservation may come at a cost to power generation as some of the emission control technologies (e.g. FGD facilities) are quite energy-intensive. Depending on the specific desulphurisation process, the energy consumption of a FGD facility could be anywhere from 0.5 to 2 % of total electricity generation (Yang 2008). The majority of thermal generation units in China use the wet limestone-gypsum FGD technology which normally uses up 1.5–2 % of total

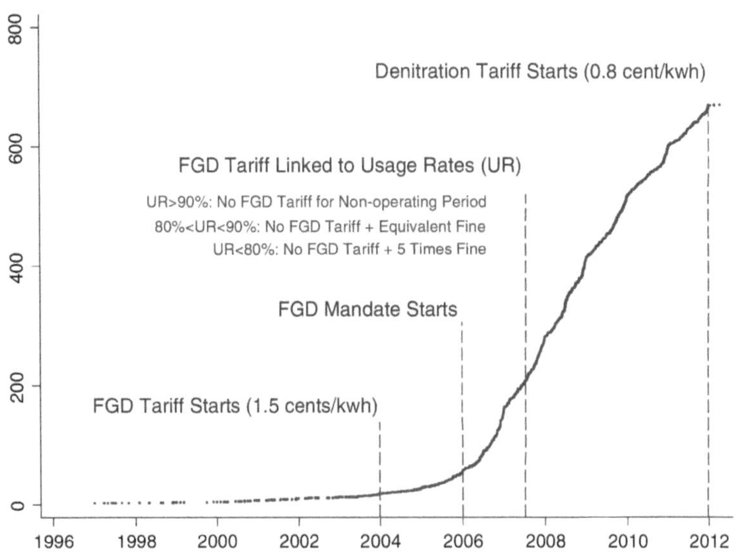

Fig. 3.7 Accumulated commissioned FGD capacity (GigaWatts). *Vertical axis* represents accumulated FGD capacity in gigawatts. *Source* Ma/Zhao (2015)

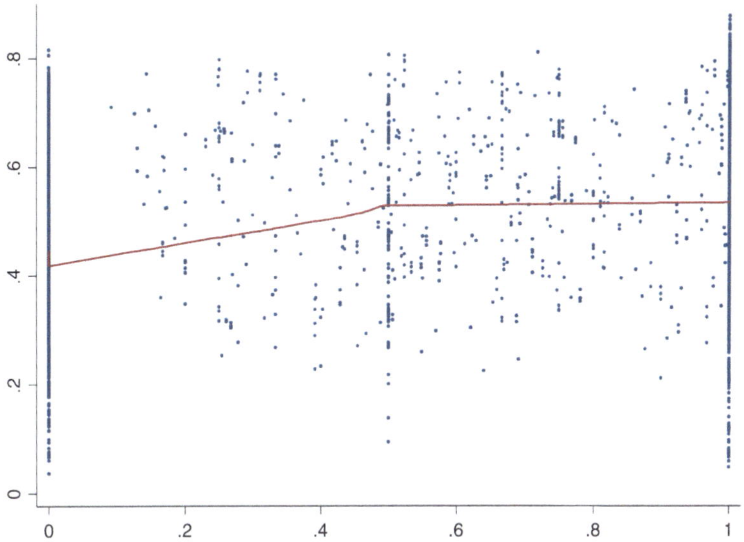

Fig. 3.8 TSE versus % desulphurised capacity. *Vertical axis* represents TSE scores and *horizontal axis* represents the proportion of total generation capacity that has been equipped with FGD facilities. *Source* Author's calculation

electricity generation. Given that the output variable is measured by net generation, running FGD facilities would increase captive use and reduce a power plant's operational efficiency.

On the other hand, it is also possible that if power plants with older units using outdated generation technology foresee a major technology upgrade or replacement in the near future, they would mostly likely delay compliance to the FGD mandate to avoid repetitive investment (in FGD). From a mid- to long-term efficiency perspective, the regulatory authority is likely to approve such a strategy and practice. As a result, FGD facilities are more often than not installed on new units with advanced generation technologies. Figure 3.8 seems to support this latter hypothesis. To fully understand the impact of emission control on the operational efficiency, it is necessary to have information on specific generation technologies. Alternatively, joint estimation of environmental and operational performance can be conducted if emission data is available.

3.5 Integrated Assessment

Economic modellers and policymakers typically calibrate future projections based on historical trends. Sensible calibration would need sound understanding of historical trajectories, and in particular, whether and to what extent the observed trend represents short-term fluctuations or even one-off changes which are less relevant for long-term projections. In the case of power plants' efficiency, part of the short-term fluctuations are due to load changes. Reifschneider/Stevenson (1991) found that departures from an efficiency frontier may reflect the systematic effect of conditions that contribute to inefficiency. Factors such as demand-induced growth rate in electricity supply or the demand constrained capacity utilisation may limit the ability of the utility to attain the frontier.[3] Because electricity cannot be conveniently stored, generation facilities follow the load across demand cycles. Although total gross capacity can be adjusted in the long run—either by retiring outdated units or installing new units—varying demand is largely met by adjusting capacity utilisation of existing units. In a less competitive market, the regulator may deviate from the optimal economic dispatch to meet other regulatory needs, e.g. to control local air quality or to achieve a fair allocation of generating hours among plants which consequentially affect a power plant's capacity utilisation. As Fig. 3.4 shows, the utilisation factors in China's power sector have mostly followed the economic cycle, which peaked in 2004 when the demand for electricity was high and supply was short. The tension between electricity supply and demand has been relieved since 2004. These demand fluctuations would have also contributed to the observed efficiency variations in the past years.

In light of these aforementioned issues (changes in technical characteristics, demand change, commissioning/decommissioning), Ma and Zhao conducted an integrated analysis of observed efficiency gain during the 1997–2010 period. They confirmed that the 2002 unbundling reform had significantly boosted the operational efficiency of China's power plants; however, the contribution was much less than previously thought. In fact, using two very different estimation techniques, their analysis shows consistent evidence that the retrofitting mandates—SUSM and PBQSP in particular—have made a contribution to the observed efficiency gain that is at least as big as that of the unbundling reform. The results have important implications for predictions of future efficiency improvement as well as in policy making. The fact that a large proportion of the efficiency improvement observed in the last decade or so is due to a number of technological mandates, implies that future efficiency gain would be much less pronounced than has been observed in the recent past. Technology mandates can only take us so far and associated efficiency gains are typically one-off dividends. Continuous efficiency improvement would have to come from a competitive market environment or other policy stimuli.

[3]There are several reasons why a plant may have a capacity utilisation factor lower than 100 %: (1) a unit may be out of service or operating at reduced output for part of the time due to equipment failures or regular maintenance; (2) output is curtailed because the electricity is not needed (e.g. lower demand); and (3) generators choose to reduce output or even shut down because the price of electricity is too low to make generation economical.

3.6 Challenges and Opportunities

Despite the Chinese government's great policy effort to refurbish its conventional energy supply and its ambitious goal for substantial penetration of renewable energy, the power sector still faces severe challenges. For the foreseeable future, fossil fuels, coal in particular, will remain dominant in China's energy supply. This is not simply because China has very abundant coal resource. The existing stock of technology also plays an important role. Although the large-scale retrofitting process has created a new fleet of generation units that are much more efficient than the old ones, and the government has banned new coal-fired plants in selected regions, China's electricity supply, to a large extent, is locked into fossil-fuel-fired generation for the next few decades due to the typical long life-cycle of these units.

A second challenge is related to continuous efficiency improvements for these fossil-fuel-fired generation units. It has been a great achievement to have substantially boosted the generation efficiency of China's power sector within a relatively short period of time. This achievement has been a result of several policy incentives. To a large extent, it is due to the large-scale retrofitting process mandated in the past few years. Without further technology upgrades, such improvements in the future will most likely be one-off. Efficiency improvement in the future will mostly likely be much less pronounced.

Following the SUSM and PBQSP, the government introduced yet another program for power plants to upgrade and retrofit with a focus on improving the heat rate of generation and upgrading capacity for combined heat and power (CHP) supply. The program encourages participation by units: (1) with a heat rate of 5 g of standard coal equivalent per kWh higher than the average; (2) with a capacity greater than 100 mW but less than 1,000 mW; (3) that have been in operation for over two years; (4) with an industrial heating supply capacity greater than 70 ton/h/unit or a space heating supply capacity over 2.4 million square meters; and (5) with an estimated annual energy saving greater than 7,500 tons of standard coal equivalent. Participating units will receive a number of favourable policy supports, including: (1) favourable approval for generation capacity expansion; (2) state (and possibly local) subsidies; (3) loans with favourable terms; and (4) favourable grid dispatch order. These new initiatives may boost power plants' efficiency to another level; however, the actual impact remains unclear as the new initiative is unlike the previous mandates, and only a voluntary program.

Support for energy and environmental conservation from both the private and public sectors is essential for continued technology upgrades. However, concerns have been raised about insufficient funding support from the governments and lack of private incentives. The government has released its 2013 government budget where conservation and environmental protection was the only major budget item to see a cut. This aroused quite a debate on whether China is underfunding its war on pollution due to economic pressure. It is hoped that the funding gap will somehow be filled by the private sector; however, a precondition for private investment is the right incentives and proper market environment. This is actually a great opportunity for China's power sector.

Technology mandates may be effective but not necessarily efficient. In principle, a competitive market will sort out the most cost-effective solutions such that we do not rely on the government to pick the winners. In addition, a competitive market provides dynamic incentives for continuous improvement while power plants generally have no strong incentives to over-comply with technology mandates. The 2002 reform formally articulated four aims: (1) unbundling of generation from the grid; (2) unbundling of main and auxiliary businesses; (3) unbundling of transmission from distribution; and (4) creating a competitive wholesale market. The first two elements have been accomplished. The reform in 2002 also aimed to gradually develop a competitive regional or even national wholesale electricity market where power plants bid to enter the market and gain grid-access priority according to economic efficiency and merit order. The competitive bidding would benefit more efficient generating units and thus create an ongoing incentive for all plants to improve productivity. A competitive market will provide the greatest opportunity for China's power sector.

Appendix: Data Sources

The main dataset in this chapter—in addition to Ma (2012), Zhao/Ma (2013), and Ma/Zhao (2015)—is from the Chinese regulatory authority that collects the most authoritative statistical data on China's electricity generation. The dataset has information on generation, consumption of various fuels, load factors, captive use, transmission losses, ownership, vintage, etc. for all existing hydro, thermal and renewable power plants that has a combined capacity over 6,000 kW.

There are thousands of power plants in the original dataset; however, for the purpose of this study, we focused on thermal power plants which include coal-, oil- and gas-fired plants, and left out hydro plants. In addition, we restricted the analysis to thermal power plants with a minimum combined capacity of over 100,000 kW in any year of the study period since matching unit information from other sources for smaller plants is much more difficult. Figure 3.9 presents the total number of power plants and the proportion of combined sample capacity to China's overall thermal capacity in each year. As is shown in Fig. 3.9, even this restricted final sample still includes the majority thermal generation capacity in China's electricity sector. As such, the results of the analyses using this data can be viewed as representative of the whole sector.

Although the dataset does provide information for each generation unit in some years, in most cases it only reports aggregated plant-level data. Information on the nameplate capacity, commissioning date, and decommissioning date of each individual unit, commissioning date of each piece of FGD equipment and the total number of generation units are collected from a number of other government documents released by the Ministry of Environmental Protection and National

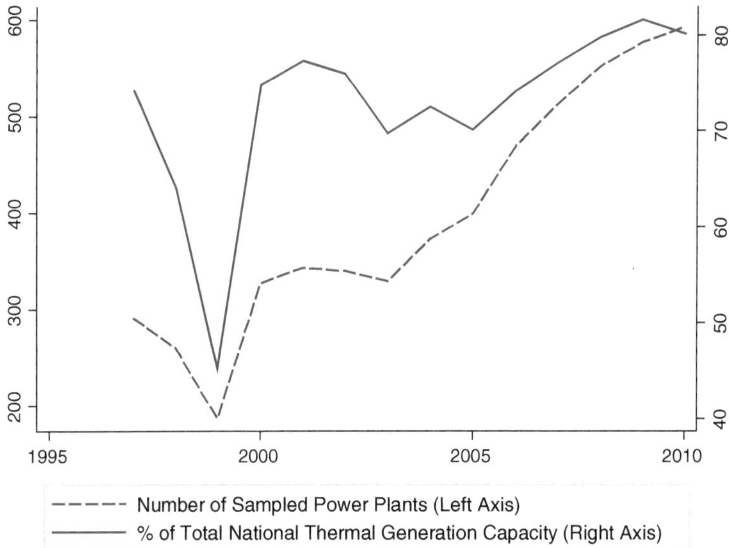

Fig. 3.9 Sample coverage. The smaller number of power plants and lower capacity coverage in 1998 and 1999 are due to missing information on a number of key variables for a large number of power plants. *Source* Ma/Zhao (2015)

Development and Reform Commission—in particular, *A List of Running Desulfurization Facilities on Coal-fired Units* (MEP 2011, 2012) and *A List of Closed Small Thermal Units* (NDRC 1999–2010, various issues). The dataset is further supplemented with information collected from the websites of individual power plants especially in cases where a unit is not equipped with an FGD facility.

Acknowledgments This project received financial support from the Australian Research Council (ARC Discovery Grant) and the Research Collaboration Award at the University of Western Australia. I thank session attendants of the 23rd Annual Conference of the Association of Chinese Economics Studies Australia (2011), the 35th Annual Conference of the International Association for Energy Economics (2012), the Theo Murphy Frontier of Sciences Conference at the Australian Academy of Sciences (2012), and the Energy Studies Institute Conference at the National University of Singapore (2013) for many valuable comments. All remaining errors are mine.

References

Aldhous, P. (2005) "China's Burning Ambition". *Nature* 435: 1152–4.

Boden, T.A., G. Marland and R.J. Andres (2010) "Global, Regional, and National Fossil-Fuel CO2 Emissions". Carbon Dioxide Information Analysis Center, Oak Ridge National Laboratory, U.S. Department of Energy, Oak Ridge, TN, USA. doi 10.3334/CDIAC/00001_V2010.

BP (2013) *BP Statistical Review of World Energy.*

Cao, Jing, R. Garbaccio and Mun S. Ho (2009) "China's 11th Five-Year Plan and the Environment: Reducing SO_2 Emissions". *Review of Environmental Economics and Policy* 3, 2: 231–50.

Du, L.M., J. Mao and J.C. Shi (2009) "Assessing the Impact of Regulatory Reforms on China's Electricity Generation Industry". *Energy Policy* 37: 712–20.

Du, L.M., Y.N He and J.Y. Yan (2013) "The Effects of Electricity Reforms on Productivity and Efficiency of China's Fossil-Fired Power Plants: An Empirical Analysis". *Energy Economics* 40: 804–12.

Gao, H. and J.V. Biesebroeck (2014) "Effects of Deregulation and Vertical Unbundling on the Performance of China's Electricity Generation Sector". *Journal of Industrial Economics* 62: 41–76.

Hiebert, L.D. (2002) "The Determinants of the Cost Efficiency of Electric Generating Plants: A Stochastic Frontier Approach". *Southern Economic Journal* 68, 4: 935–46.

Hughes, V. (2012) "Where There's Smoke". *Nature* 489: S18–20.

Joskow, P. and R. Schmalensee (1987) "The Performance of Coal-Burning Electric Generating Units in the United States: 1960–1980". *Journal of Applied Econometrics* 2: 85–109.

Lam P.L. and A. Shiu (2004) "Efficiency and Productivity of China's Thermal Power Generation". *Review of Industrial Organization* 24: 73–93.

Ma, C.B. (2012) "China's Green Economy". Presentation at Theo Murphy Frontier of Science Conference, Australian Academy of Science.

Ma, C.B. and L.N. He (2008) "From State Monopoly to Renewable Portfolio: Restructuring China's Electric Utility". *Energy Policy* 36: 1697–711.

Ma, C.B. and X.L. Zhao (2015) "Electricity Market Restructuring and Technology Mandates: Plant-Level Evidence for China's Changing Operational Efficiency from 1997 to 2010". *Energy Economics* 47: 227–37.

Ministry of Environmental Protection (MEP) (2011; 2012) *A List of Running Desulfurization Facilities on Coal-Fired Units*. Beijing: MEP.

National Development and Reform Commission (NDRC) (1999–2010). *A List of Closed Small Thermal Units,* various issues. Beijing: NDRC.

Pollitt, M.G. (1995) *Ownership and Performance in Electric Utilities*. Oxford: Oxford University Press.

Reifschneider, D. and R. Stevenson (1991) "Systematic Departures from the Frontier: A Framework for the Analysis of Firm Inefficiency". *International Economic Review* 32: 715–23.

Smith, S.J., J. van Aardenne, Z. Klimont et al. (2011) "Anthropogenic Sulfur Dioxide Emissions, 1850–2005: National and Regional Data Set by Source Category". Version 2.86. Palisades, NY: NASA Socioeconomic Data and Applications Center (SEDAC). http://sedac.ciesin.columbia.edu/data/set/haso2-anthro-sulfur-dioxide-emissions-1850-2005-v2-86 on March 25, 2014.

Tollefson, J. (2008) "Stoking the Fire". *Nature* 454: 388–92.

Yang, Q.Y. (2008) "Summarization on Desulfurization Technology of Thermal Power Plant". *Environmental Protection Sciences* 34, 3: 8–11 [in Chinese].

Yang, H.L. and M. Pollitt (2009) "Incorporating Both Undesirable Outputs and Uncontrollable Variables into DEA: The Performance of Chinese Coal-Fired Power Plants". *European Journal of Operational Research* 197: 1095–105.

Zhao, X.L. and C.B. Ma (2013) "Deregulation, Vertical Unbundling and the Performance of China's Large Coal-Fired Power Plants". *Energy Economics* 40: 474–83.

Chapter 4
Efficiency Improvement in China's Energy-Intensive Industries and Their Contributions to Carbon Emission Reduction Targets

Libo Wu

Abstract China is presently undergoing an intensive industrialisation. As the world's largest emerging economy, energy efficiency is expected to play a critical role in China's ever-rising demand for energy. Based on factual overviews and numerical analysis, this chapter presents an in-depth investigation into the effectiveness of policies announced or implemented in recent decades targeted at energy conservation in energy-intensive manufacturing sectors. It highlights nine energy-intensive sectors that achieved major improvements in their energy technology efficiency efforts. Under the umbrella of the 11th Five-Year Plan, the success of these sectors' energy-saving efforts reflects the effectiveness of China's energy conservation policies. The Chinese government has introduced various measures to reduce the road transport sector's demand for energy and its greenhouse gas (GHG) emissions: by implementing fuel economy standards, and by promoting advanced energy-efficient vehicles and alternative fuels. Coal-based energy-saving technologies, especially industrial furnace technologies, are critical for China's near and medium-term energy-saving goals, which include the improvement of the direct technical efficiency for power generation, iron and steel production, non-ferrous products production, and non-metal mineral products production. In the long run, renewable energy development and expanding the railway transport system are the most effective ways to reduce China's energy use and GHG emissions. Its road vehicles are projected to consume 370–520 million metric tons of oil and produce 1.6–2.0 billion tons of GHGs by 2050. Fuel economy standards could reduce oil consumption and GHGs by 34–35 %.

Keywords Energy efficiency · Energy intensive sectors · Transportation sector · China

---

Dr. Libo Wu, Professor, School of Economics, Fudan University, China;
Email: wulibo@fudan.edu.cn.

B. Su and E. Thomson (eds.), *China's Energy Efficiency and Conservation*,
SpringerBriefs in Environment, Security, Development and Peace 30,
DOI 10.1007/978-981-10-0737-8_4

4.1 Introduction

In the last two decades of the 20th century, the intensity of China's energy use fell rapidly at a rate unparalleled by any other country at a similar stage of industrialisation. Internally, the dramatic expansion of labour-intensive industries allowed for the substitution of cheap labour to energy inputs and provided market-based incentives for energy conservation. Decomposition analysis focusing on Chinese industrial energy use has showed such substitutions (Zhang 2003; Wu et al. 2005, 2006). Externally, the Chinese government has implemented various policies and programmes to monitor industrial energy use, disseminate energy conservation information and services, stimulate energy technology renewal, promote investment in innovation and improve energy management capabilities (Sinton et al. 1998, 1999; Sinton/Fridley 2000; Wang et al. 1995). Energy consumption in China has accelerated since 2001 and surpassed GDP growth in 2003 and 2004. As a result, the long-lasting decline of energy intensity ended—the annual decreasing rate of energy intensity was –4.02 % in the 1980s and –6.30 % in the 1990s. However, this reversed to 0.36 % in the first five years of the 21st century (see Fig. 4.1).

Concerned about energy security, ever-rising energy costs, climate risks and environmental degradation, the Chinese government enacted the 20 % energy intensity reduction target in the 11th FYP. Following this challenging goal, central

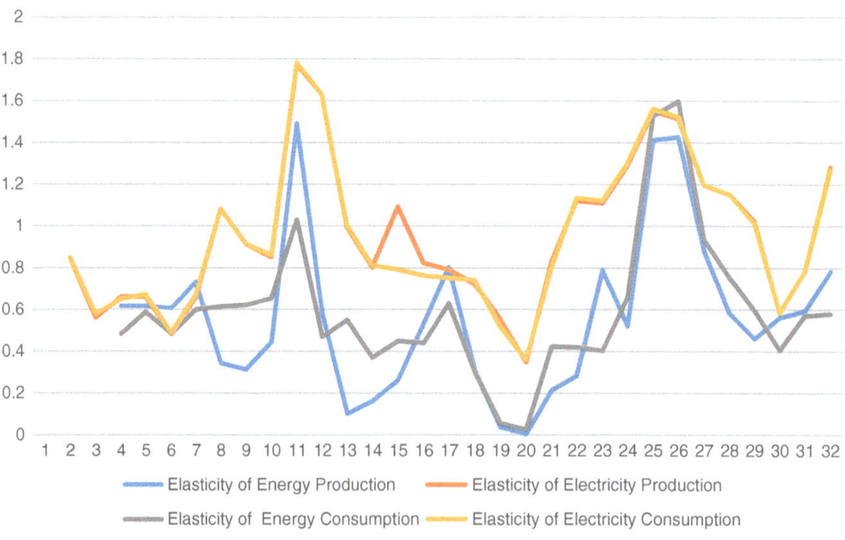

Fig. 4.1 Elasticity of energy production and consumption. *Source* China Energy Statistical Year Book 2010

and regional governments took various political, technical, industrial and fiscal measures to facilitate the energy conservation activities of stakeholders. From 2006 to 2010, China's energy use per unit of economic output fell by 19.06 % (State Council 2011). In the 12th FYP covering 2011–2015, the Chinese government announced that energy intensity is to decline by 16 % and carbon intensity by 18 %. All these targets are further disaggregated to each province by the National Development and Reform Commission (NDRC).

Along with these mandatory energy conservation and carbon emission reduction targets, energy-intensive sectors were forced to get rid of a number of energy technologies and in the meantime, utilise new energy-saving technologies recommended by the national government. Given the intensive industrialisation and motorisation taking place in China, can these command-and-control policies work effectively to completely change the ever-rising energy consumption pattern? This chapter will addresses this question by clarifying the energy consumption scale as well as the structural and efficiency dynamics in the Chinese industrial sector, and examine the relevant policy implications Furthermore, by introducing a new portfolio of energy-saving technologies into the I-Dream model, this chapter also provides some simulation results with respect to the contribution of these technologies to low-carbon emissions reduction targets in the long run.

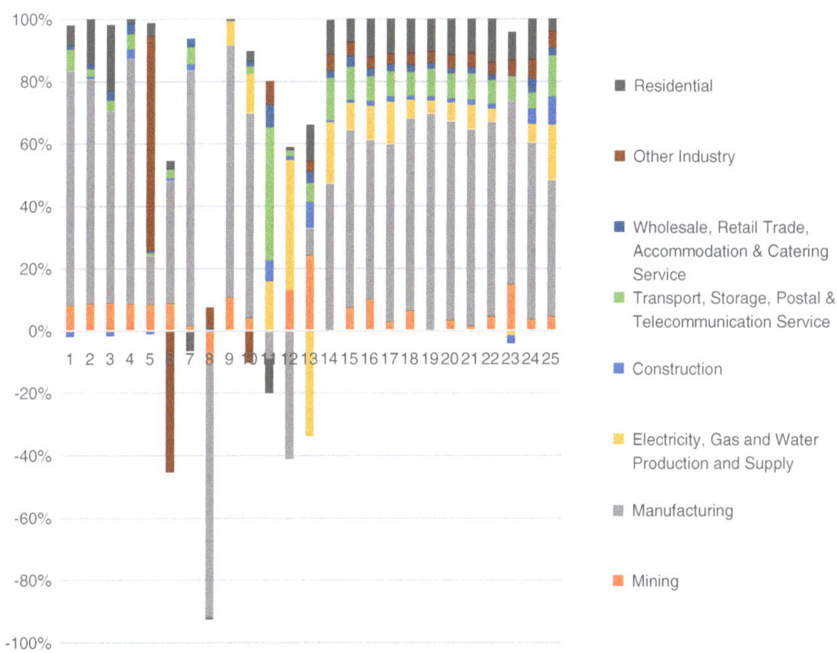

Fig. 4.2 Sectoral contribution to the annual incremental energy use in China. *Source* Author's calculation

The accelerated pace of industrialisation, urbanisation and motorisation accounts for most of China's energy use. Figure 4.2 shows that during the past decade, the contribution of the manufacturing sectors to incremental energy use experienced an inverted-U curve. It was 55.31 % in 2000, peaked at 67.95 % in 2004, and then declined to 43.2 % in 2010. The transportation sector, together with the power generation sector, accounted for a greater portion of the incremental energy use over the same period.

Is the above structural transformation mainly due to the command-and-control policy since the 11th FYP? Alternatively, is there an intrinsic mechanism also playing a significant role in driving the change? How does one rate the energy efficiency performance of the Chinese industrial sector, especially the energy-intensive sectors before and after strong political interventions? What are the likely contributions of energy conservation technologies to long-term energy conservation and carbon emissions reductions in China? This chapter employs numerical analysis to determine the relative importance of the driving forces of industrial energy use, and discusses the policies that China is implementing to reduce energy and their effects on the industrial energy use.

4.2 Energy Efficiency Analysis in China's Manufacturing Sector

4.2.1 Overview of Energy Conservation Policies in the Recent Decades

In the late 1990s, energy consumption in China's manufacturing industries declined greatly, contributing to the slowdown of total energy use (see Fig. 4.1). Since 2000, however, energy use in manufacturing industries rebounded to become the dominant forces driving the rising energy demand. In 2005, nine energy-intensive sectors accounted for 96.2 % of total manufacturing industrial energy use, 78.2 % of total industrial energy use and 55.9 % of total final energy consumption, but their gross output was just 44.0 % of the manufacturing total and 41.4 % of the industrial total. Energy-intensive sectors were also the main sources of environmental pollutants, such as SO_x, NO_x and suspended particulates. These facts pointed to the urgent need to slow the scale of expansion of energy-intensive sectors, to improve their energy efficiency and to adjust product production methods.

In the 11th FYP, the Chinese government announced a comprehensive energy intensity reduction target and put the energy-intensive sectors at the centre of energy conservation. However, this did not hinder the rapid expansion of energy-intensive sectors in the initial stages of its implementation. Regional goals for economic growth and large sunk costs made it difficult to slow the overall scale of energy-intensive sectors in the short term. Efficiency improvement, by reducing energy input per unit of output, became more critical to counteract expansion (Zhou et al. 2010; Price et al. 2011). In order to strengthen the effectiveness of the energy conservation movement, the State Council, the National Development and Reform Commission (NDRC) and other responsible ministries and administrations announced and implemented various policies. The following section examines all these policies and measures over the 11th FYP period and the beginning year of the 12th FYP.

It is clear that the governance focus was on the energy-intensive sectors, with special attention to the 1,000 large-scale enterprises.

4.2.1.1 Strengthening the Energy Conservation Technology Diffusion

The Chinese government has organised a large amount of specific funding for energy conservation and emissions reduction. The funding system mainly serves as a public subsidy to improve the cost efficiency of the 10 key energy-saving projects. The NDRC published the *10 Key Energy Saving Projects Guidelines* and the Ministry of Science and Technology published the *National Technology Policy Outline on Energy Conservation* to instruct the technology transformation and innovation directions. Any energy conservation project approved by the authorised expert committee can receive RMB 200–300 per ton of standard coal for the energy-saving units. In the *National Technology Policy Outline for Comprehensive Resource Utilisation*, technologies aimed at improving life cycle energy efficiency are introduced, and systematic change presently carries more importance as compared to end-of-pipeline management. Specifically, 3R (reduce, reuse and recycle) principles are strongly recommended for application in the production processes; hence less energy is consumed and less pollutants are emitted when producing the same amount for outputs.

4.2.1.2 Strengthening the Capability-Building for Key Enterprises, Regional Government and Professional Third Parties

Capability-building policies have been widely implemented over the past decades to develop the energy auditing, monitoring and statistical system. In 2005, the rule of reporting energy use per unit of GDP was established. The National Administration of Statistics and NDRC announced the energy conservation performance of each province regularly. During the 11th FYP, energy management capability-building projects were mainly focussed on the 1,000 key energy-intensive enterprises. A regular reporting system is now well developed among these enterprises (Dai/Zhou 2008; Zhou et al. 2010). In 2010, policies aimed at the state-owned key enterprises and medium and small-scale enterprises were also announced.

4.2.1.3 Phasing Out Technologies and Industries with Energy-Exhausting and Outdated Production Capacity

In addition to the technology innovation and capability-building, the Chinese government is taking strong measures to phase out outdated production capacities in energy-intensive sectors. Small-scale plants with outdated production capacity (from power generation, coal-mining and processing, coking production, ferroalloy production, calcium carbide production, iron and steel production, cement and plate glass production, fibre and clothing production) are on the agenda to be removed. While there are many critics of intervention in the market, there are some people in regional governments who are supportive of this policy. It is reported that over the period of the 11th FYP, China phased out 76,825 MW of power generation capacity, 120 million tonnes of pig iron production capacity, 72 million tons of steel production capacity and 370 million tons of cement production capacity.

How do we evaluate the effects of the above policies with respect to their effectiveness and efficiency? Looking at the energy use situation in 2010, we find that the gross output contribution of energy-intensive sectors to total manufacturing and industrial energy use declined by 2.2 % compared to 2005 levels. However, their share in total manufacturing energy use, total industrial energy use and total final energy consumption declined slightly by 0.2–0.5 %. Further analysis is needed to clarify the relative importance of the factors driving the inelastic energy demand of energy-intensive sectors and to determine the correct policy orientation and rationale of the 12th FYP target.

4.3 Scale, Structure and Efficiency Dynamics of Energy Use in China's Manufacturing Sector

4.3.1 *Brief Model Description*

By applying the basic logarithmic mean divisia index (LMDI) decomposition techniques developed by Wu et al. (2005), this chapter analyses the energy consumption of China's manufacturing sectors. With our focus on energy-intensive sectors, we include such sectors and take the other manufacturing sectors as a whole. We thus put forward this model to examine concerns about the dominant energy-intensive sectors users, and define the following KAYA type equation as the basis for decomposition model:

$$\text{Energy Consumption} = \text{Gross Output} \times \frac{\text{Added Value}}{\text{Gross Output}} \times \frac{\text{Physical Output}}{\text{Added Value}} \times \frac{\text{Energy Consumption}}{\text{Physical Output}} \tag{4.1}$$

Then, the final format of the decomposition model can be depicted as the following equation:

$$\frac{E_T}{E_0} = \exp\sum_{i=1}^{10}\tilde{\omega}_i \ln\frac{SF_T}{SF_0} \times \exp\sum_{i=1}^{10}\tilde{\omega}_i \ln\frac{ECEF_T}{ECEF_0} \times \exp\sum_{i=1}^{10}\tilde{\omega}_i \ln\frac{AVRF_T}{AVRF_0} \times \exp\sum_{i=1}^{10}\tilde{\omega}_i \ln\frac{ENEF_T}{ENEF_0} \quad (4.2)$$

We use the time series data from 1994 to 2010, while sectoral energy consumption, gross output (current price) and added value (current price) data until 2007 can be taken directly from the *National Statistical Yearbook* and *Energy Statistical Yearbook* (NBS 1993–2011). We use the yearly sectoral PPI index as a deflator to derive the constant price. Since 2007, the Administration of Statistics has only released the year-on-year growth rate and annual compound growth rate for sectoral added value. We estimate the constant price added value of 2008–2010 by taking 2007 as the base year.

4.3.2 Model Result and Discussion

4.3.2.1 The Relative Importance of Various Factors to Energy Consumption Change in the Manufacturing Sectors

The decomposition model indicates that during the past decades, especially during the 11th FYP, energy consumption in China's manufacturing sectors has decelerated and even stabilised (see Fig. 4.3). This transition is not mainly due to the

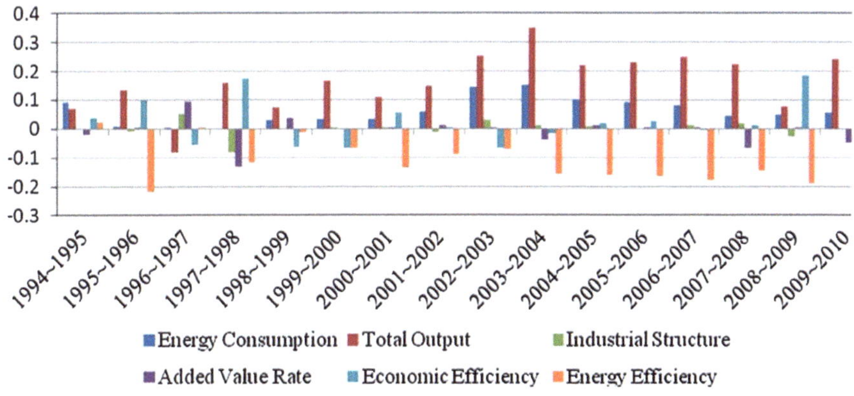

Fig. 4.3 Relative importance of various factors to energy consumption in manufacturing sector. *Source* Author's calculation

slowdown in economic growth, but because of energy efficiency improvements. In fact, with the exception of the crisis year of 2008, total output expansion has always played a dominant role in driving up energy use. Energy efficiency improvement acts as an opposite force pushing down energy use. However, since their scale could not match that of output expansion at the beginning of this century, the net outcome of industrial energy use soared substantially. The imbalance was reduced in 2003, and energy efficiency improvements have been a considerable source of dampening expansion since that time.

The Plan for National Economic and Social Development plays a central role in guiding the direction of policymaking in China. A comparison across the 9th, 10th and 11th FYP (Fig. 4.3) indicates the importance of energy efficiency improvements as they became more pronounced in the 11th FYP; while the scale of change for economic output did not alter for each five-year period. Therefore, this study illustrates that policies towards energy efficiency improvement in the industrial sectors, especially the energy-intensive sectors in the 11th FYP did not reverse the internal economic growth trends but added substantially to overall energy efficiency improvement.

4.3.2.2 Relative Importance of Factors to Energy Consumption Change in Various Energy-Intensive Sectors

Figure 4.4 illustrates some interesting findings generated by the sectoral analyses. For most of the energy-intensive sector, excluding the power generation sector, the total energy efficiency improvement was spectacular in the 11th FYP. Furthermore, energy technology innovation, the phasing out of outdated small-scale production capacity, and improved managerial policies contributed to considerable efficiency improvement, which effectively counteracted expansion. In particular, energy efficiency improvement in the raw chemical product industry, smelting and processing of ferrous industry, non-chemical mineral products industry, and paper-making industry successfully outperformed the economic growth that brought about noticeable shrinkage in overall energy use.

However, it is noted that the pace of energy efficiency improvement in the power generation sector is becoming less. This is clearly indicated in Fig. 4.4. During the past decade, the gross coal consumption rate for fossil-fired power plants declined by 11.85 %, but the absolute amount is still 8.9 % higher than the average level of Japan in 2010. The structure of the power generation source is the main factor hampering the overall thermal power efficiency improvement. In 2008, coal consumption still accounted for 89.3 % of China power generation, while in Japan, only 27.6 % of thermal power was based on coal with 23.6 % from natural gas and 30.7 % from nuclear power. This means that while there is still some technical space for overall efficiency improvement in fossil-fired power plants, more efforts are needed to facilitate less dependence on fossil fuels.

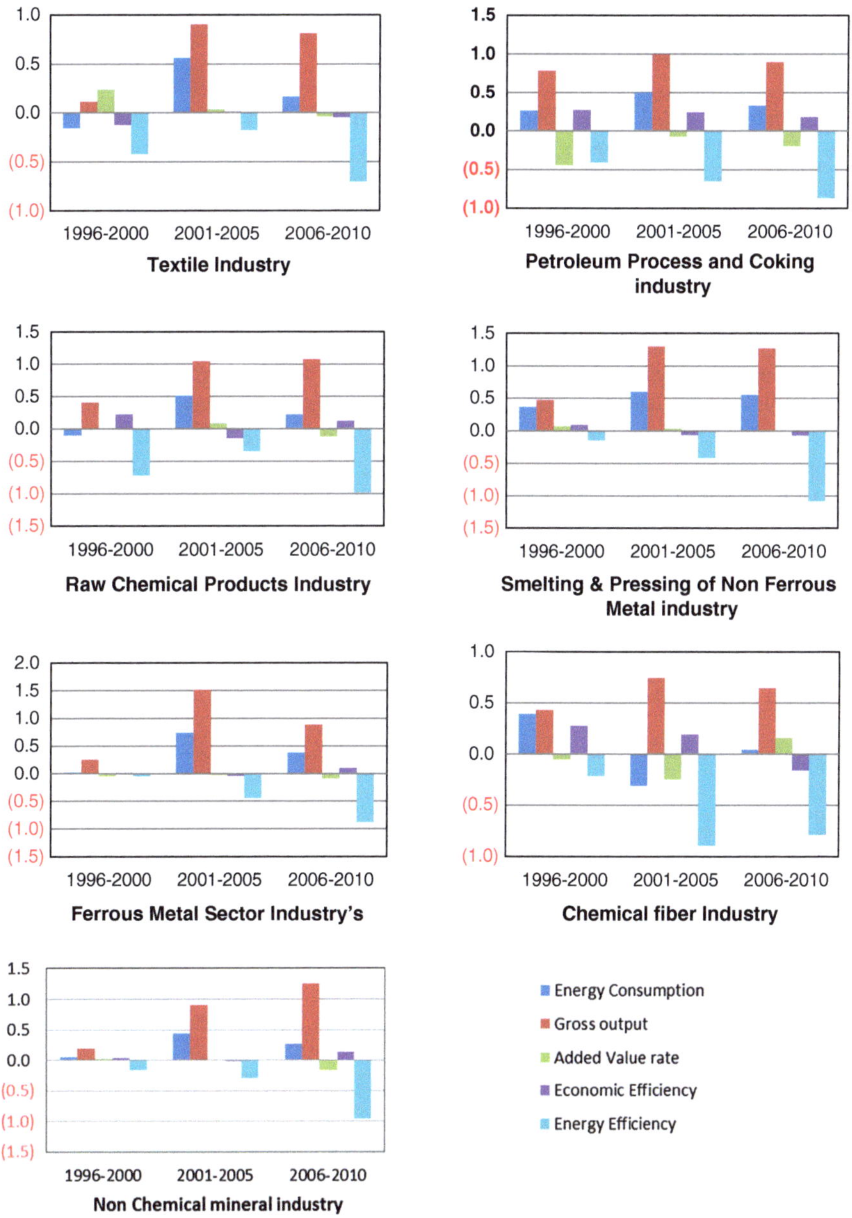

Fig. 4.4 Change of energy use and its driving forces in energy-intensive sectors. *Source* Author's calculation

4.3.2.3 Relative Importance of Various Sectors to Energy Consumption Change

As indicated in Fig. 4.5, the abnormal increase in the manufacturing sector's energy use over the 11th FYP period was driven manly by the smelting and processing of the ferrous metal sector, non-metal mineral products sector, and raw chemical materials and product sector.

Chinese production of steel and pig iron, cement, synthetic ammonia and alkaline increased rapidly during the past decade. China has been the largest producer of pig iron in the world this century and its consumption of it continues to increase. In the past decade, pig iron production rose by 342 %, whereas the global production increased by less than 50 %. Therefore, China's production accounted for 46.39 % of the world's total in 2010 and was about two times the sum of Japan, US, Germany, Russia and India. Given the fact that the comparable energy consumption for steel in China was still higher than that in Japan by 1.4 % in 2000 and 10.95 % in 2010, China's dramatically expanding ferrous industry should make energy conservation its key priority. Over the past decade, China contributed more than 90 % to the world incremental cement production. The energy efficiency of cement production in China was lower than Japan's by 44 % in 2000 and 22.8 % in 2010, respectively. Therefore, the cement production sector should also make more efforts to reduce its energy use.

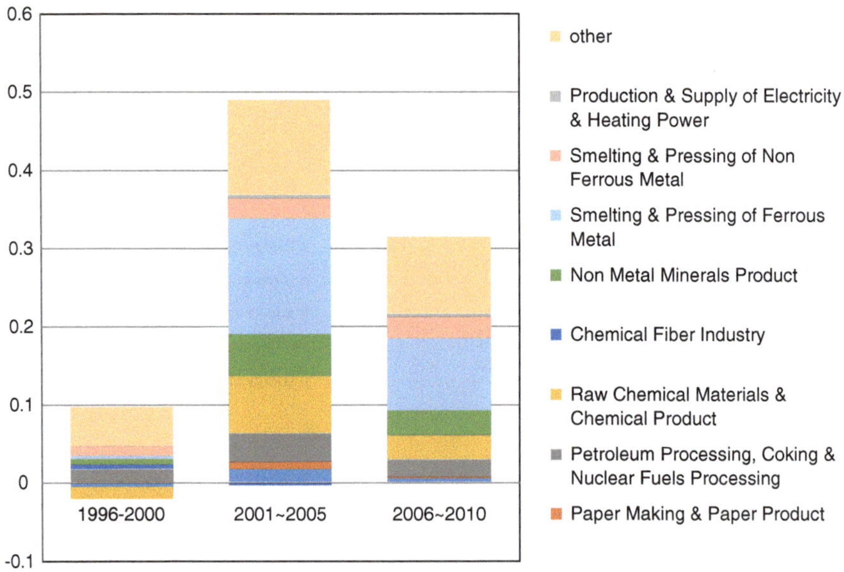

Fig. 4.5 Relative importance of various sectors to total manufacturing energy use. *Source* Author's calculation

4.4 Modelling Technological Development in Energy-Intensive Sectors in a Low-Carbon Economy: A Shanghai Case Study

4.4.1 Model Introduction

In recent years, the low-carbon transformation of China has been further accelerated by domestic environmental degradation, energy security concerns and rising costs of energy use. Energy-intensive sectors are playing central roles in this process by utilising more energy-saving and emissions reduction technologies. Furthermore, the Chinese government regularly provides recommendations for technologies to guide the direction of innovation and remove inefficient production capacities.

This chapter examines how energy-intensive sectors can choose an optimised technological roadmap under different scenarios of energy conservation and emissions reduction targets. It will go on to illustrate the scenario analysis results of the Shanghai model: differing from previous studies, this model modifies the commonly utilised energy technology portfolio of the TIMEs model by introducing a series of energy conservation technologies recommended by the Chinese government. Therefore, the analysis results are capable of capturing energy technology innovations since the 11th FYP in all of the energy-intensive sectors.

4.4.1.1 Development of the Reference Energy System

The basic structure of our Shanghai TIMES model can be referred to in Wu (2010). This model establishes a reference energy system (RES) containing 10 energy processing technologies, 15 power and heat generation technologies, 100 industrial technologies, 9 residential technologies, 9 service technologies and 31 transport technologies. In order to clarify the competitiveness of newly recommended technologies by the Chinese government, this model is further modified to add 21 key conservation technologies in the "ten major energy conservation projects implementation planning" (NDRC 2011). Since this chapter's focus is on energy-intensive sectors, these 21 technologies are identified as new energy-use technologies in iron and steel, construction materials, petro-chemical products and non-ferrous metal products industries. The following figure illustrates some special modification to the reference energy system (RES) by introducing detailed technology choices into energy-intensive sectors regarding their energy use (Fig. 4.6).

4.4.1.2 Parameters and Main Assumptions

The second step consists of inputting data for all the parameters of energy commodities and technologies into the model. These parameters are composed of

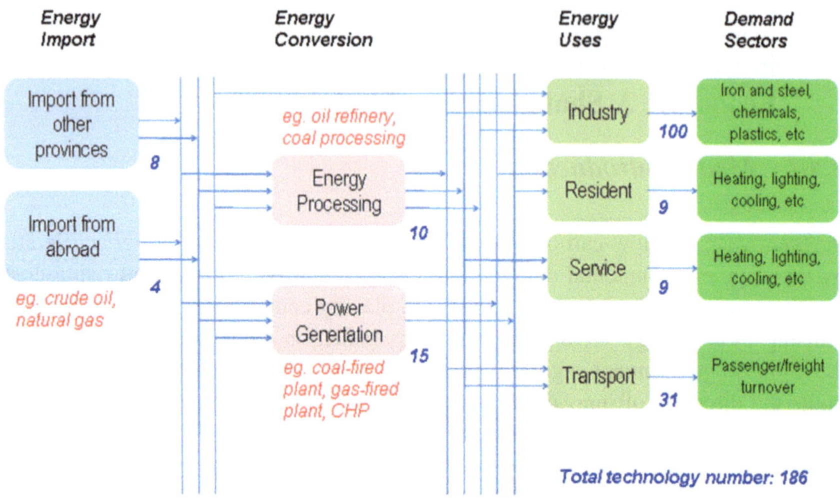

Fig. 4.6 RES of Shanghai TIMES model. *Source* TIMES model results

resource supply curves, demand projections, dynamically evolving technology costs and other relevant parameters.

The resource supply curves are based on current energy prices and IEA forecasts (see Table 4.1). This model assumes the domestic energy prices in China will change in line with the global energy prices after 2020.

Demand projections are based on historical energy demand in Shanghai from 2000–09. The compounded annual growth rates (CAGR) of different energy demands are computed across all demand sectors from 2005–09 to estimate the energy demand in 2010. As the growth rate of all demand devices will not remain at such a high level in the future, we assume the CAGR will decrease by half every 10 years in each sector. By estimating the CAGR of energy demands during 2010–20, 2020–30, 2030–40 and 2040–50, the model can extrapolate the energy demands of all the milestone years from 2010–50 (Table 4.2).

Technology costs comprise the investment cost, the fixed O&M[1] cost and the variable O&M cost of each energy technology. As TIMES is a technology-rich model which minimises the total cost of the energy system, these cost data are very important to modelling. The investment cost data comes mainly from official statistics and our field surveys, while the O&M cost data is estimated by setting an appropriate ratio compared to the investment cost data.

[1]O&M cost represents the operation and maintenance cost of an energy technology. For example, in a coal-fired power plant, the fixed O&M cost includes the salaries of employees and the repair fees of power equipment, while the variable O&M cost implies the cost of electricity and some fuel oil associated with the use of the power plant.

Table 4.1 Assumptions on energy import prices

	Unit	2000	2010	2020	2030	2040	2050
Import prices of energy commodities from other provinces in China							
Hard coal	RMB yuan/tonne	188.78	720.00	881.95	1043.85	1200.43	1200.43
Crude oil	RMB yuan/tonne	959.44	3000.00	7256.01	10102.02	12122.42	12122.42
Natural gas	RMB yuan/m^3	0.78	2.80	4.41	6.13	7.97	7.97
Diesel	RMB yuan/tonne	2306.77	7295.20	12211.89	17001.73	20402.08	20402.08
Kerosene	RMB yuan/tonne	1087.03	3400.00	5754.69	8011.84	9614.20	9614.20
Gasoline	RMB yuan/tonne	2484.65	7964.00	13153.57	18312.77	21975.32	21975.32
Fuel oil	RMB yuan/tonne	1397.62	4300.00	7398.88	10300.93	12361.12	12361.12
LPG	RMB yuan/tonne	1664.50	5300.00	8811.73	12267.94	14721.53	14721.53
Hard coal	US dollar/tonne	33.65	128.81	157.21	186.07	213.98	213.98
Import prices of energy commodities from abroad							
Crude oil	US dollar/barrel	28.00	90.00	148.23	206.37	247.64	247.64
Natural gas	US dollar/mmbtu	3.87	13.72	19.64	27.28	35.46	35.46
LNG	US dollar/mmbtu	4.73	13.63	19.56	27.16	35.31	35.31

Source CEIC, IEA and TIMES model results

Table 4.2 Total demand projections in different sectors

	Unit	2000	2010	2020	2030	2040	2050
Industrial sector	PJ	515	1,400	2,145	2,851	3,360	3,671
Residential sector	PJ	147	307	808	1,328	1,708	1,938
Service sector	PJ	73	155	423	706	916	1,044
Transport sector							
Passenger turnover	Mn-passenger-km	23,472	111,141	311,454	528,343	690,546	790,169
Freight turnover	Mn-freight-km	662,000	1,507,706	2,328,162	2,899,922	3,238,431	3,422,749

Source Shanghai Bureau of Statistics and TIMES model results

Other relevant parameters include conversion efficiency, technology life, residual capacity and emission factor. These parameters are estimated by referring to relevant OECD data and making some adjustments according to the technological conditions in China.

4.4.1.3 Scenario Settings

In this chapter, one base scenario as a benchmark scenario and three low-carbon scenarios (namely LC1, LC2 and LC3) are set to analyse the technological development pattern of the economy.

Base scenario is a business-as-usual scenario which includes legislated policy measures as of 2010. However, no future policies or measures would be enacted to reduce energy use or CO_2 emissions.

LC1 scenario is a low-carbon scenario involving a low level of emissions restrictions. At the Copenhagen Meeting in 2009, the Chinese government promised to reduce carbon emissions per unit of GDP by 40–45 % in 2020 from 2005 levels. Shanghai, a vanguard city for a low-carbon economy, is expected to do better than the whole country in achieving these emissions targets. It is estimated that Shanghai will reduce its carbon emissions per unit of GDP by 50 % in 2020 from its 2005 level. Additionally, with structural upgrades to its economy, Shanghai will likely reduce its carbon emissions per unit of GDP by 65 % in 2030 from the 2005 base. China should ideally assume absolute quantitative control of its carbon emissions by 2040, while Shanghai is expected to retain its CO_2 emissions at its 2030 levels. By 2050, Shanghai is likely to further reduce its CO_2 emissions by 10 % from its 2040 level.

LC2 scenario is a low-carbon scenario involving a medium level of emissions restrictions. Shanghai is expected to decrease its carbon emissions per unit of GDP by 55 % in 2020 from the 2005 base. Then, by 2030, Shanghai is likely to quickly assume absolute quantitative control of carbon emissions by limiting CO_2 emissions growth to no more than 40 % from the 2020 base. By 2040, Shanghai is expected to retain its CO_2 emissions to 2030 levels, and by 2050, it appears set to further reduce its CO_2 emissions by 10 % from the 2040 base.

LC3 scenario is a low-carbon scenario involving a high level of emissions restrictions. Shanghai is likely to reduce its carbon emissions per unit of GDP by 60 % in 2020 from the 2005 base. Then, by 2030, Shanghai is expected to assume quantitative control of its carbon emissions by limiting CO_2 emissions growth to no more than 30 % from the 2020 base. By 2040, Shanghai will likely retain its CO_2 emissions at its 2030 level, and by 2050, it is expected to further reduce its CO_2 emissions by 10 % from the 2040 base.

4.4.1.4 Core Scenario Results

In Fig. 4.7, the bars in four different colours represent the amount of CO_2 emissions from 2000–50 in four respective scenarios, and the lines denote the development of CO_2 emissions per unit of GDP.

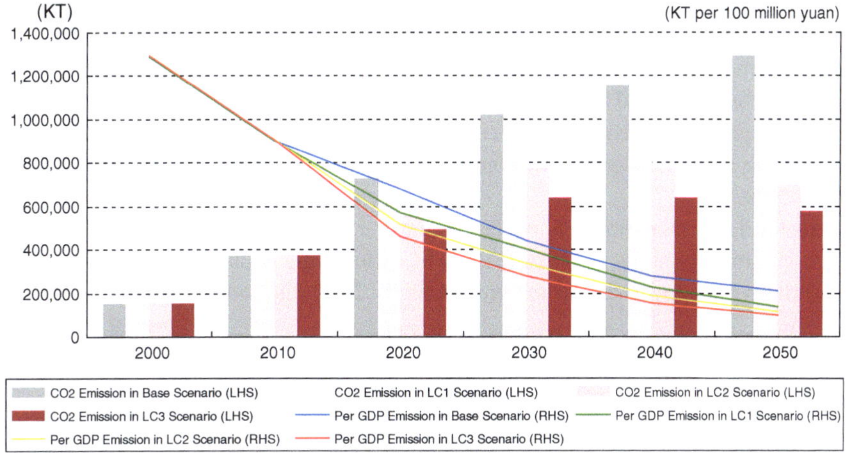

Fig. 4.7 Shanghai's CO_2 emissions in different scenarios, 2000–50. *Source* TIMES model results

4.4.1.5 Overall Results

As shown, in the base scenario, Shanghai's CO_2 emissions will increase dramatically from 153,634 kt in 2000 to 1,292,529 kt in 2050. However, as the economy develops, the carbon emissions per unit of GDP will decline rapidly from 32.20 kt per 100 million yuan in 2000 to only 5.26 kt per 100 million yuan in 2050. In the LC1, LC2 and LC3 scenarios, CO_2 emissions in 2050 will drop sharply from 1,292,529 to 836,127 kt, 697,120 kt and 575,400 kt, respectively. Meanwhile, the carbon emissions per unit of GDP will also decline to 3.40 kt, 2.84 kt and 2.34 kt per 100 million yuan, respectively.

In the Shanghai TIMES model, CO_2 emissions come from six sectors, namely: power generation, energy processing, industrial, residential, services and transport. Breaking down the sources of carbon emissions across these sectors, we find that most of the CO_2 emissions are produced by the industrial and power generation sectors.

In the base scenario, the power generation sector contributed 56 % of CO_2 emissions in the whole energy system of 2000. This is expected to grow steadily to 70 % by 2030 and then remain constant until 2050. The industrial sector produces the second largest proportion of carbon emissions, which is likely to decline from 36 % in 2000 to 23 % by 2050.

In the LC1 scenario, the proportion of carbon emissions in the power generation sector shrinks a little during the 2020–50 period compared with that of the base scenario. This is because a more stringent carbon restriction requires this sector to use more decarbonised technologies such as wind power or solar power. The proportion of the industrial sector keeps a premium of 2–4 ppt higher than that in the Base Scenario during the 2020–40 period. However, it suddenly drops by 6 ppt in 2050 after the economy assumes absolute quantitative control of carbon emissions in 2040. Others, including the residential, services and transport sectors, mildly expand their proportions after 2010. By comparing Figs. 4.8 and 4.9, one can conclude that the power generation sector has the largest potential in carbon reduction at the initial stage of decarbonisation. But when the economy assumes absolute quantitative control over carbon emissions, a significant change will occur within the industrial sector. Section 4.3 will further examine what is happening in this sector.

In the LC2 scenario, it is observed that the power generation sector produces high carbon emissions by 2020 (Fig. 4.10). But as the economy assumes absolute quantitative control over carbon emissions by 2030, the industrial sector becomes

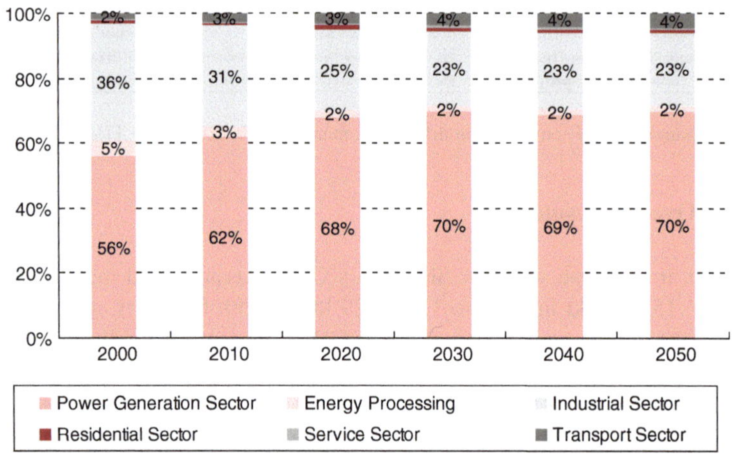

Fig. 4.8 CO_2 emissions breakdown in the base scenario. *Source* TIMES model results

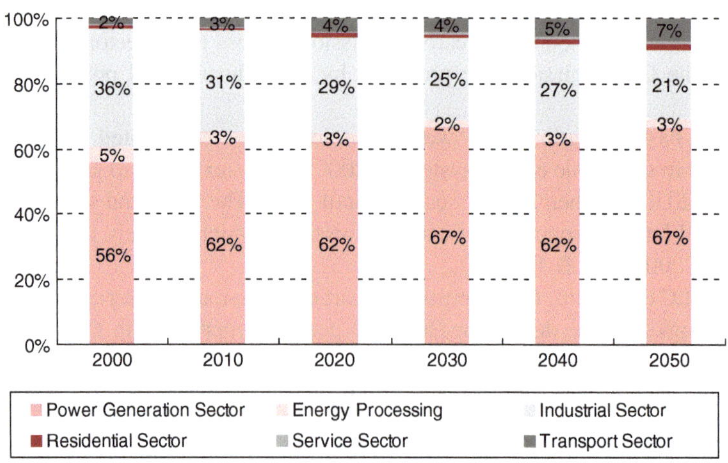

Fig. 4.9 CO_2 emissions breakdown in the LC1 scenario. *Source* TIMES model results

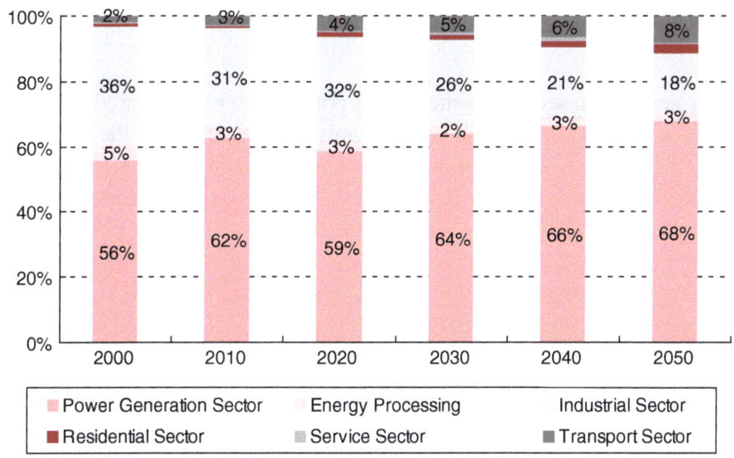

Fig. 4.10 CO_2 emissions breakdown in the LC2 scenario. *Source* TIMES model results

the vanguard sector to decarbonise as the proportion of this sector's emissions drops significantly from 32 % in 2020 to 18 % in 2050. However, other sectors have almost doubled their proportion of emissions during the 2010–50 period.

In the LC3 scenario, the carbon emissions sources have witnessed a more significant structural change as the power generation and industrial sectors produce higher proportions of emissions (Fig. 4.11). For example, most strikingly, the transport sector's emissions increase from 3 % in 2010 to 10 % in 2050. The expansion of the other three sectors' emissions is mainly due to the relatively lower contribution of their emissions as a whole and little room for technological decarbonisation at the current rate of innovation progress.

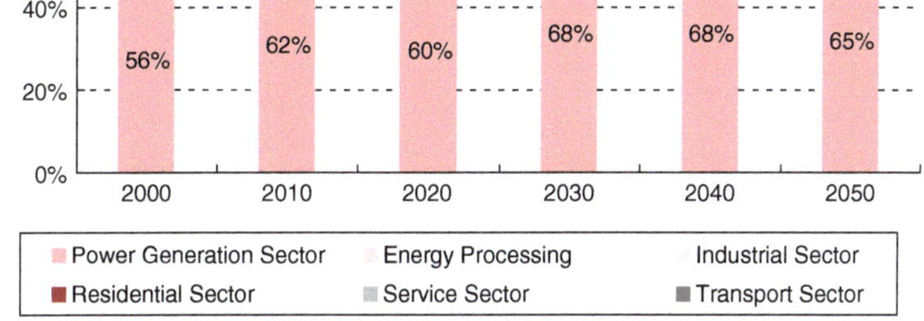

Fig. 4.11 CO_2 emissions breakdown in LC3 scenario. *Source* TIMES model results

4.4.2 Power Generation Sector

The power generation sector has the largest potential to reduce carbon emissions as it contributes the largest proportion of emissions among all the sectors. With improvements to government policies concerning carbon restrictions, the power generation sector currently uses more green energy to produce electricity.

In the base scenario, the coal-fired power plants dominate the power generation sector, with 86.1 % of electricity produced by coal technologies in 2000 (Fig. 4.12). Traditional coal technology will be gradually replaced by coal IGCC after 2020. By 2050, all coal technologies are expected to generate 96.7 % of China's electricity. As the availability of fossil fuels become scarcer, such a development pattern is not sustainable.

In the LC1 scenario, coal technologies will stop growing after 2030, and start to decrease by 2040 as the Chinese economy assumes quantitative control of carbon emissions (Fig. 4.13). By 2050, coal technologies will likely produce only 47.0 % of electricity while wind and CHP will contribute to 38.6 and 14.3 %, respectively, of electricity in the power generation sector. This means that wind power technology will be applied on a large scale after 2030. As Shanghai is geographically located near the sea, we expect that offshore wind power technology will be the most prospective technology to support the decarbonisation progress of the economy.

In the LC2 scenario, the power generation sector will be further decarbonised as coal technologies will account for only 31.5 % of electricity produced (Fig. 4.14). Wind and CHP will in turn contribute to 41.6 and 19.1 % of electricity, respectively. Notably, the economy will start to use solar power after 2040, and solar power generation will likely account for 8.0 % of electricity produced by 2050.

In the LC3 scenario, renewable technologies will produce more than half of the electricity produced in the power generation sector (Fig. 4.15). The application of such a high proportion of renewable technologies will create supply shocks to the

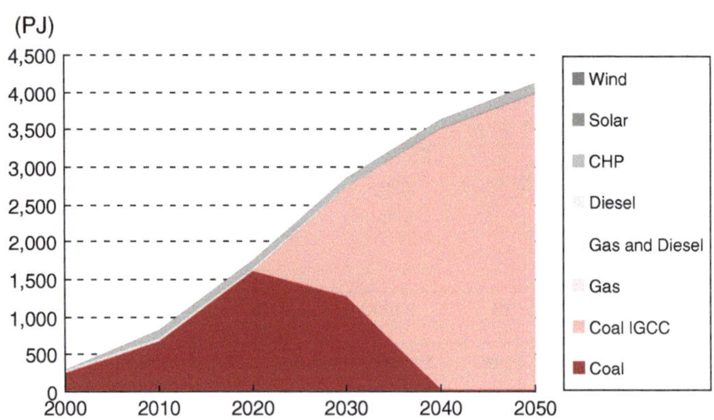

Fig. 4.12 Energy consumption of the power sector in base scenario. *Source* TIMES model results

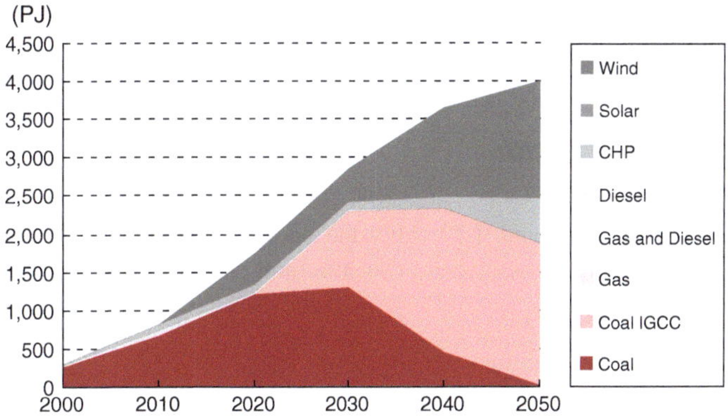

Fig. 4.13 Energy consumption of the power sector in LC1 scenario. *Source* TIMES model results

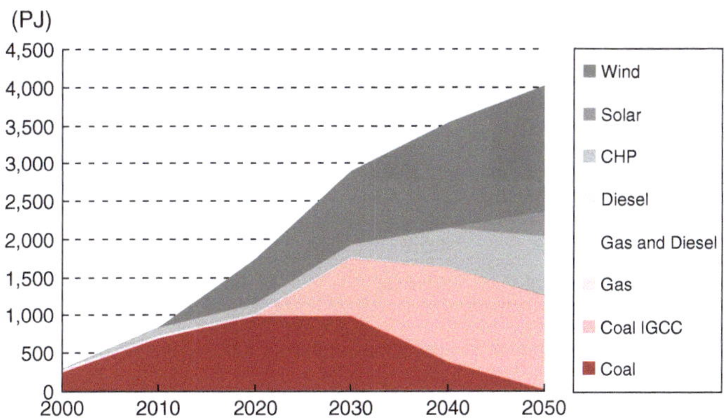

Fig. 4.14 Energy consumption of the power sector in LC2 scenario. *Source* TIMES model results

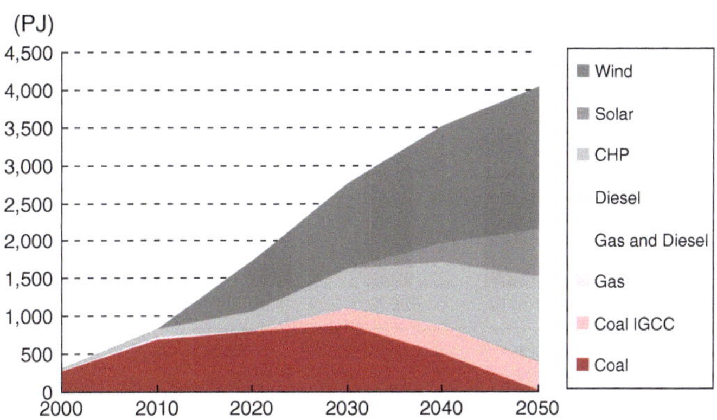

Fig. 4.15 Energy consumption of the power sector in LC3 scenario. *Source* TIMES model results

grid system, and thus hasten the urgency to develop energy storage technology to offset the uncertainty of renewable technologies.

4.4.3 Industrial Sector

Figures 4.16, 4.17, 4.18 and 4.19 show the energy consumption structure of the industrial sector in all four scenarios. The structural changes occurring in this sector

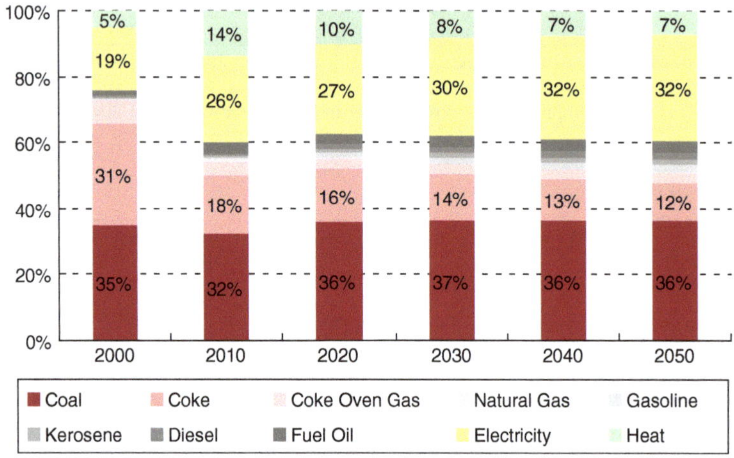

Fig. 4.16 Energy consumption of the industrial sector in base scenario. *Source* TIMES model results

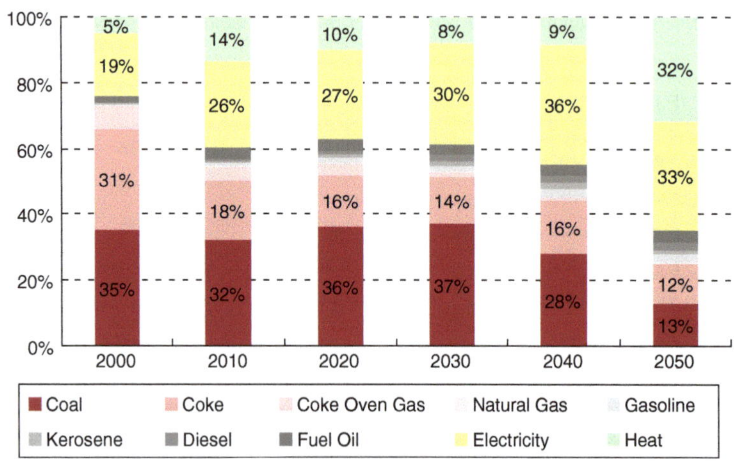

Fig. 4.17 Energy consumption of the industrial sector in LC1 scenario. *Source* TIMES model results

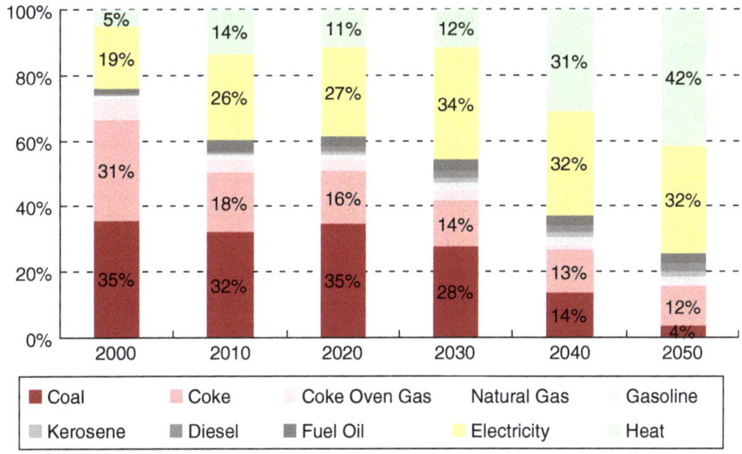

Fig. 4.18 Energy consumption of the industrial sector in the LC2 scenario. *Source* TIMES model results

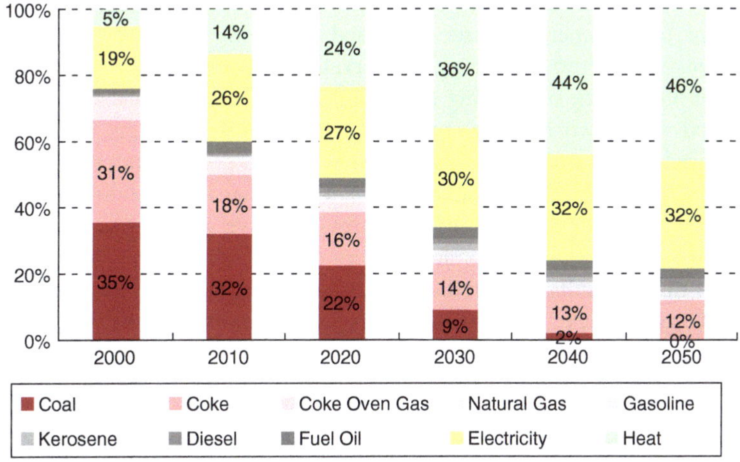

Fig. 4.19 Energy consumption of the industrial sector in the LC3 scenario. *Source* TIMES model results

reflect a distinct pattern of decarbonisation. As carbon restrictions become more and more stringent across the low-carbon scenarios, the proportion of coal technologies in 2050 drops sharply from 36 % in the base scenario to 13, 4 and 0 % in LC1, LC2 and LC3 scenarios, respectively. Another notable change is that heat technologies account for 32, 42 and 46 % of all energy consumption in LC1, LC2 and LC3 scenarios, respectively, while in the base scenario only 7 % of energy consumption is produced from heat technologies.

As discussed in Sect. 4.1, when the Chinese economy assumes absolute quantitative control of carbon emissions, the industrial sector will reduce its carbon emissions significantly. This is done by eliminating more coal technologies and using cleaner technologies such as electricity and heat in the industrial sector. Electricity technologies are already expected to contribute to about one-third of energy consumption in the base scenario by 2050. But why will the proportion of heat technologies grow so fast under the low-carbon scenarios? First, only electricity and heat technologies do not produce carbon emissions. Second, electricity is largely produced by coal-fired power plants, while heat is produced by CHP using a large proportion of natural gas which is less carbon-intensive than coal. In addition, using more renewables to produce electricity is more costly than using current CHP technology to produce heat, thus making heat technologies more favourable than electricity technologies in efforts to meet stringent carbon emission restrictions.

4.5 Conclusions

In the early years of the 12th FYP, the Chinese government released a series of policies for further energy conservation and emissions reduction. Following the 1,000 Intensive Enterprises Energy Conservation Project, 10,000 enterprises were included in the energy monitoring and auditing system. A decline in international exports and slowdown of economic growth in domestic markets have also dramatically reduced the demand for energy-intensive goods, and resulted in the problem of over-production in these sectors. The updated list detailing the phasing out of outdated production capacity in energy-intensive sectors was thus approved despite great pressure for economic recovery (MOIIT 2012).

In light of the results and predictions discussed throughout this chapter, continuity and consistency in China's energy conservation scheme can be safely expected for the next decade. Industrial structure upgrades and technology innovation are regarded as the two underpinnings of the country's energy conservation efforts to achieve systematic optimisation of the energy system in the long term.

This chapter presents a TIMES model and three low-carbon scenarios are applied to analyse the technological development of Shanghai in a low-carbon economy. Giving insights into the energy system of Shanghai, one base scenario and three low-carbon scenarios with different CO_2 emission restrictions are set to observe the kinds of changes that would take place in a low-carbon economy. By comparing the emissions structure among different sectors and analysing the energy consumption source of each sector, we reach the following conclusions which present us with a development pattern of a low-carbon economy in view of technological advancement or replacement.

Among all the sectors in Shanghai's energy system, the power generation sector has the largest potential for carbon reduction at the initial stage of decarbonisation. However, when the economy assumes absolute quantitative control of carbon emissions by the industrial sector will become a vanguard to reduce CO_2 emission by facilitating technological reforms.

The decarbonisation of the power generation sector is accomplished mainly by cutting down the proportion of coal-related technologies and using more renewable technologies. In the LC3 scenario, the most stringent low-carbon scenario, more than half of all electricity is produced by wind and solar power. Such a high proportion of renewable technologies applied will create supply shocks to the grid system. Therefore, the development of energy storage technology is urgently needed to offset the uncertainty of renewable technologies.

Significant changes have occurred in the energy consumption structure of China's industrial sector as its economy further reduces its carbon emissions. The large proportion of coal technologies used in the base scenario will gradually be replaced by more clean technologies such as electricity and heat technologies in low-carbon scenarios. It is noted that the proportion of heat technologies expands extremely fast from the LC1 to LC3 scenarios, chiefly due to the fact that using more renewables in producing electricity is more costly than using current CHP technology to produce heat.

Appendix

See Table 4.3.

Table 4.3 Energy conservation policies in China over the past decade

Measures	Category	Date effective	Stakeholders	Mandatory ratings	Sectoral overage
Energy conservation law (revision version)	National law	Apr. 2008	National People's Congress	Obligation (instructional support)	All sectors
Circular Economy Promotion Law	National law	Jan. 2009	NPC	Obligation (instructional support)	All resource-related sectors
The State Council's decision on strengthening energy conservation	National regulation	Aug. 2006	State Council	Obligation	Mainly focused on energy-intensive sectors
The State Council's announcement of a comprehensive scheme on energy conservation and emissions reduction	National regulation	June 2007	State Council	Obligation	Mainly focused on energy-intensive sectors

(continued)

Table 4.3 (continued)

Measures	Category	Date effective	Stakeholders	Mandatory ratings	Sectoral overage
The State Council's announcement that approved the statistics, monitoring and auditing systems for energy conservation and emissions reduction	National regulation	Nov. 2007	State Council	Obligation	All provinces and energy-intensive sectors, especially the 1,000 energy-intensive enterprises
The State Council's announcement of strengthening oil and electricity conservation	National regulation	Aug. 2008	State Council	Obligation	Transportation sectors, electricity sectors and other related sectors
The State Council's announcement of the 12th FYP's comprehensive scheme on energy conservation and emissions reduction	National regulation	Aug. 2011	State Council	Obligation	All sectors
Mid- and long-term planning for energy conservation	National planning	Nov. 2004	NDRC	Obligation to national and regional governments (instructional support)	All sectors, focusing on energy-intensive sectors
11th FYP	National planning	Mar. 2006	NDRC	Obligation to national and regional governments (instructional support)	All sectors, focusing on energy-intensive sectors
12th FYP	National planning	Mar. 2011	NDRC	Obligation to national and regional governments (instructional support)	All sectors, focusing on energy-intensive sectors and more energy users

(continued)

Table 4.3 (continued)

Measures	Category	Date effective	Stakeholders	Mandatory ratings	Sectoral overage
Decision on the implementation of 10 key energy conservation projects in the 11th FYP	Ministry policy	July 2006	NDRC; Ministry of Science and Technology; Ministry of Finance; Ministry of Housing And Urban-Rural Development; General Administration of Quality Supervision, Inspection, Quarantine; Ministry of Environmental Protection	Partial obligation to national and regional governments (instructional support)	Mainly focused on energy-intensive sectors
Announcement restricting the export of energy, pollution and resource-intensive products	Ministry policy	Dec. 2005	NDRC; MOF; Ministry of Commerce; MOLR; General Administration of Customs General Administration of Taxation; MOEP	Partial obligation to national and regional governments (instructional support)	Mainly focused on energy-intensive sectors
Announcement to establish the rule of reporting energy use for per unit of GDP	Ministry policy	Dec. 2005	NDRC; Office of the National Energy Leading Group; National Bureau of Statistics	Obligation	All provinces and energy intensive sectors
Guidance on strengthening the energy-saving and emissions reduction of medium and small-scale enterprises	Ministry policy	Apr. 2010	MOIIT	Partial obligation to national and regional governments (instructional support)	Energy-intensive sectors

(continued)

Table 4.3 (continued)

Measures	Category	Date effective	Stakeholders	Mandatory ratings	Sectoral overage
Tentative measurement on monitoring the energy conservation and emissions reduction of state-owned key enterprises	Ministry regulation	Mar. 2010	State Owned Assets Supervision and Administration Commission of the State Council	Obligation to state-owned key enterprises	State-owned key enterprises
Announcement of strengthening the energy-saving assessment and auditing for fixed assets investment	Ministry regulation	Mar. 2010	MOIIT	Obligation to national and regional governments (instructional support)	All sectors, mainly energy-intensive sectors
National technology policy outline on energy conservation	National technology planning	Feb. 2007	NDRC; MOST;	Instructional support	Mainly focused on energy-intensive sectors
National Technology policy outline for comprehensive resource utilisation	National technology planning	July 2010	NDRC; MOST; Ministry of Industry and Information Technology; MOHURD; Ministry of Land and Resources	Instructional support	All resource-related sectors
Action plan for energy conservation in 1,000 energy-intensive enterprises	Ministry policy	2006	NDRC; National Energy Agency; NBS; GMQSI; SOASA		1,000 energy-intensive enterprises

Source Author's summary

References

Alfstad, T. (2005) "Development of a Least Cost Energy Supply Model for the SADC Region". Master thesis, University of Cape Town.

Chen, W. (2005) "The Costs of Mitigating Carbon Emissions in China: Findings from China MARKAL-MACRO Modeling". *Energy Policy* 33, 7: 885–96.

Chen, W. and Z. Wu (2001) "Use MARKAL Model to Study the Future Sustainable Energy Development Strategy in China". *Journal of Tsinghua University (Science and Technology)* 41, 12: 103–6 (in Chinese).

Chen, W. and Z. Wu (2002) "China's Future Policies to Control SO_2 and CO_2". *Journal of Tsinghua University (Science and Technology)* 42, 10: 1320–3. (in Chinese).

Chen, W., P. Gao and J. He (2004) "Use MARKAL-MACRO Model to Study the Impact of Carbon Emission on Chinese Energy System". *Journal of Tsinghua University (Science and Technology)* 44, 3: 342–6. (in Chinese).

Chen, W., Z. Wu, J. He et al. (2007) "Carbon Emission Control Strategies for China: A Comparative Study with Partial and General Equilibrium Versions of the China MARKAL Model". *Energy* 32, 1: 59–72.

Chen, W., J. Liu, L. Ma et al. (2009) "Role for Carbon Capture and Storage in China". *Energy Procedia* 1, 1: 4209–16.

Dagoumas, A.S. and T.S. Barker (2010) "Pathways to a Low-Carbon Economy for the UK with the Macro-Econometric E3MG Model". *Energy Policy* 38, 6: 3067–77.

Delaquil, P, W. Chen and E.D. Larson (2003) "Modeling China's Energy Future". *Energy for Sustainable Development* 7, 4: 40–56.

Gielen, D. and C. Chen (2001) "The CO_2 Emission Reduction Benefits of Chinese Energy Policies and Environmental Policies: A Case Study for Shanghai, period 1995–2020". *Ecological Economics* 39: 257–70.

Hainoun, A., M. Seif Aldin and S. Almoustafa (2010) "Formulating an Optimal Long-Term Energy Supply Strategy for Syria using MESSAGE Model". *Energy Policy* 38: 1701–14.

Kannan, R. (2009) "Uncertainties in Key Low Carbon Power Generation Technologies: Implication for UK Decarbonisation Targets". *Applied Energy* 86, 10: 1873–86.

Makela, J. (2000) "Development of an Energy System Model of the Nordic Electricity Production System". Master thesis, Helsinki University of Technology, 2000.

Manne, A., R. Mendelsohn and R. Richels (1995) "MERGE: A Model for Evaluating Regional and Global Effects of GHG Reduction Policies". *Energy Policy* 23, 1: 17–34.

Sands, R.D. (2004) "Dynamics of Carbon Abatement in the Second Generation Model". *Energy Economics* 26, 4: 721–38.

Song, H.-J., S. Lee, S. Maken, S.-W. Ahn et al. (2007) "Environmental and Economic Assessment of the Chemical Absorption Process in Korea using the LEAP Model". *Energy Policy* 35, 10: 5109–16.

Wu, Z., P. Delaquil, E.D. Larson et al. (2001) "Future Implications of China's Energy-Technology Choices: Summary of a Report to the Working Group on Energy Strategies and Technologies". *Energy for Sustainable Development* 5, 4: 19–31.

Zhang, Z.X. (2003) "Why Did the Energy Intensity Fall in China's Industrial Sector in the 1990s? The Relative Importance of Structural Change and Intensity Change". *Energy Economics* 25, 6: 625–38.

Chapter 5
Potential of CO_2 Abatement Resulting from Energy Efficiency Improvements in China's Iron and Steel Industry

Ying Fan, Lei Zhu and Yuan Li

Abstract The iron and steel industry is one of the world's as well as China's largest energy CO_2 emission sources. We calculated the cost of CO_2 abatement (CCA), and give the marginal abatement cost curve of the main process of iron and steel production. Based on this, we analyse the cost-effectiveness of CO_2 abatement technologies. Also, we define a two-country (home and foreign), two-goods (home goods and foreign goods) partial equilibrium model to simulate China's iron and steel industry, and analyse the influence of CO_2 price and free allocation to the production, price, income, profit and total emissions of China's iron and steel industry. We found that a carbon market would increase the abatement cost and then increase the domestic price, but a reasonable free allocation could offset the profit loss partly, so it would not result in a huge profit loss to the industry.

Keywords Abatement potential · Carbon market · Partial equilibrium model · Iron and steel industry

5.1 Introduction

The iron and steel manufacturing industry is one of the world's largest energy-consuming sectors, accounting for more than 5 % of the world's annual energy demand. It is also one of the largest CO_2 emission sources in the world. In 2010, China's crude steel production reached 627 million tons, accounted for 44 % of global crude steel production, and its average growth rate has been 7 % over for the past ten years. The iron and steel industry is also one of China's largest CO_2

Dr. Ying Fan, Professor, Center for Energy and Environmental Policy Research, Institute of Policy and Management, Chinese Academy of Sciences; Email: yfan@casipm.ac.cn.

Dr. Lei Zhu, Associate Professor, Center for Energy and Environmental Policy Research, Institute of Policy and Management, Chinese Academy of Sciences.

Ms. Yuan Li, Ph.D. Student, Center for Energy and Environmental Policy Research, Institute of Policy and Management, Chinese Academy of Sciences.

B. Su and E. Thomson (eds.), *China's Energy Efficiency and Conservation*,
SpringerBriefs in Environment, Security, Development and Peace 30,
DOI 10.1007/978-981-10-0737-8_5

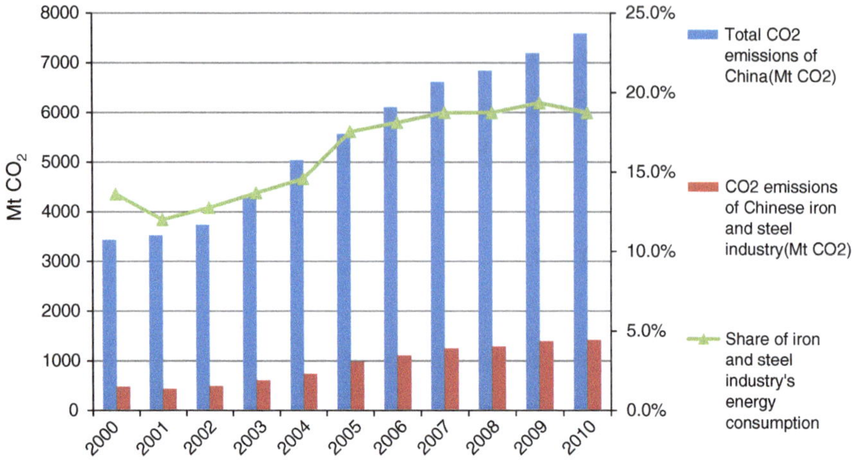

Fig. 5.1 CO_2 emissions of China's iron and steel industry. *Source* China Iron and Steel Industry Yearbook, 2011

emission sectors, accounting for more than 17 % of China's total CO_2 emissions since 2006 (Fig. 5.1).

China's iron and steel industry has two disadvantages. First, the main process of China's steel production—the Blast Furnace-Basic Oxygen Furnace (BF-BOF)—consumes more energy than the Electric Arc Furnace (EAF) process, and emits more CO_2. Second, coal accounts for more than 70 % of China's energy consumption, making it a high CO_2 emission source. And with the shares of petrol and gas lower than the international advanced level, this results in the high CO_2 emissions from China's iron and steel industry.

Improving energy efficiency is the best way to achieve reduced CO_2 emissions from the iron and steel industry. To date, the Chinese government has been paying attention to the need to save energy and reduce CO_2 emissions in the iron and steel industry, but many difficulties remain as to the issue of CO_2 abatement. China's iron and steel enterprises are much decentralised and the industrial concentration is still very low. At the same time, there is a huge gap between the large-to-medium iron and steel enterprises, and the small enterprises. The former has more advanced equipment and technologies than the former. These factors make it challenging to pursue energy saving and CO_2 abatement in China's iron and steel industry.

Promoting advanced energy-saving and CO_2 abatement technologies is one of the main abatement measures for the iron and steel industry, especially advanced technologies which have good abatement effects. Because of the heterogeneity of the different enterprises, the promotion of technology varies within the specific contexts of different enterprises. That is to say, it is important to combine the abatement effect with the cost-effectiveness when promoting technologies. Especially in recent years, with weak demand for iron and steel products and the related enterprises facing a general decline in profits (indeed, some even running into deficit), the cost-effectiveness of technologies is ever more important to these iron and steel enterprises in their technology adoption decisions.

Many CO_2 abatement technologies were promoted during the 11th and 12th Five-Year-Plans. There are currently seven carbon market pilots (Beijing, Shanghai, Tianjin, Chongqing, Hubei, Guangdong and Shenzhen) in China that are exploring an emissions trading scheme. However, enterprises are concerned about the cost of CO_2 abatement technologies (CATs), as well as possible loss of their competiveness edge when joining the carbon market. Thus, we, the authors, first studied the cost of CO_2 abatement in China's iron and steel industry through so-called conserved supply curves (CSCs), and then obtained a MAC curve by fitting the CSCs. Finally, with a partial equilibrium analysis, we studied the impact of the carbon market on the iron and steel industry.

There have been studies of various sectors' strategies for energy saving and CO_2 abatement across many different countries. The main method used is the ranking of the energy-saving/CO_2 abatement technologies based on their cost of energy savings/emissions reductions, in the form of so-called conserved supply curves (CSCs) or abatement cost curves (Worrell et al. 2000; Hasanbeigi et al. 2010; Fleiter et al. 2012). There is vast heterogeneity across the iron and steel industries of different countries; for example, in terms of production process, production structure, and technology adoption. CSC is a bottom-up method used to study the cost-effectiveness of different technologies, so it can be viewed as country-specific. Such curves could also show the CO_2 abatement contribution, potential and related cost-effectiveness of various technologies and measures.

In China, studies have focused on the iron and steel industry's energy savings and CO_2 abatement. Some of them used macro-analysis (Price et al. 2012; Guo et al. 2010; Wang et al. 2007; Zhang et al. 2012), while others have paid more attention to the micro level of China's iron and steel industry (Hasanbeigi et al. 2012). Technology-based micro-level research may be more practical in terms of allowing the governments and industries to set benchmarks for the iron and steel industry. There is presently no micro-level CATs database in China's iron and steel industry, which has made it difficult for further study of sectoral emission abatement potential estimation from the bottom-up perspective. At the same time, for enterprises, this method could be more operable than the macro-research. We, the authors, therefore collected CATs widely used in the iron and steel industry. Basing this current study on the previous, we have extended this research's range to 41 CO_2 abatement technologies, including most of the energy CATs of China's iron and steel industry. In addition, after several rounds of discussions and interviews with experts and specialists from the iron and steel industry, we have significantly calibrated the data collected in the technology list so as to reflect the actual situation in China.

There were three steps in our research: first, we analysed the production process of iron and steel and listed 41 CATs. After calculating the cost of conserved energy based on the Chinese data from 2010, we obtained a CSC for China's iron and steel industry. Second, we forecasted the CO_2 abatement potential of China's iron and steel industry in 2020 and 2030 by changing the diffusion rate of technologies and the share of BOF and EAF, and we compared the change in the CSC depending on the year and the CO_2 abatement potentials in three scenarios for 2020 and 2030, respectively. Third, we fitted a MAC curve based on the CSC, and then analysed the impact on enterprises' competiveness using partial equilibrium analysis.

5.2 Methodology

We use a CSC to rank the CATs and measures according to their cost of CO_2 abatement (CCA). From this curve, we can obtain the related cost of operating and maintaining a technology, as well as the total cost of choosing a specified technology, which includes investment, operational costs, and so on. The calculation of a CCA is shown in Formula (5.1). Formula (5.1) is referred to Worrell (2001), and then we made a change in the numerator, which subtracts the cost of saved energy. As different technologies save different energy, we calculated the cost of saved specific energy and then removed it from the numerator, which could reflect different kinds of saved energy from different CATs. Of course, when we removed the cost of saved energy, zero is used as the compare line but not the averaged energy price as with Worrell (2001).

$$\text{CCA} = \frac{\text{Annuzlized Investment} + \text{Annual Change in O\&M Cost} - \text{Cost of Saved Energy}}{\text{Annual } CO_2 \text{ abatement}} \tag{5.1}$$

The calculation of annualised investment is shown in Formula (5.2).

$$\text{Annualized Investment} = \text{Capital Cost} \times \frac{\text{d}}{(1-(1+d))^{-n}} \tag{5.2}$$

In formula (5.2), d represents the discount rate, and n represents the payback time of the CATs. Although different technologies have different payback times, from an accounting perspective, we assume that all technologies have the same payback time. Enterprises prefer a short payback time and a high internal rate of return. We thus assume a unified payback time here, which is not only easy to account for, but also reflects the preference for a short payback time in the iron and steel industry.

Also, we defined a two-country (home and foreign), two-goods (home goods and foreign goods) model to simulate China's iron and steel industry. The model adopted here is an improved version of Demailly/Quirion (2008). Here we considered the prices and elasticities of home and imported goods, as well as the potential benefit of loss resulting from emission trading schemes (ETS) in the profit calculation. And based on historical data, we estimated the price elasticities of imports and exports of China's iron and steel industry.

p_h	Price of home goods that are consumed domestically
p_x	Price of home goods that are exported
p_m	Price of imported goods
q_h	Production of home goods that are consumed domestically
q_x	Production of export goods
q_m	Production of import goods
$c_e(ua)$	Abatement cost of home goods, which is a function of unitary abatement (ua) and determined by the MAC curve
p_{CO_2}	Price of CO_2, which is exogenous
u_e	Unitary emissions of the industry
FA	Free allowance that government gives to the industry
σ_x	Export price elasticity
σ_m	Import price elasticity
θ	Price elasticity of demand

Thus the profit of the iron and steel industry at t = 1 could be defined in Formula (5.3):

$$\pi(q, ua) = p_h^1 \times q_h^1 + p_x^1 \times q_x^1 - c_e(ua) \times (q_h^1 + q_x^1) + p_{CO_2} \times FA \times u_e^0 \tag{5.3}$$

The first three parts of Formula (5.3) could be defined as the profit on production. Imports could be defined as seen in Formula (5.4):

$$q_m^1 = q_m^0 \times \left(\frac{p_m^1}{p_m^0}\right)^{\sigma_m} \tag{5.4}$$

Exports could be defined in Formula (5.5):

$$q_x^1 = q_x^0 \times \left(\frac{p_x^1}{p_x^0}\right)^{\sigma_x} \tag{5.5}$$

Total demand, which is equal to total consumption, could be defined in Formula (5.6):

$$D^1 = q_h^1 + q_m^1 - q_x^1 = \left(1 - \left(1 - \frac{p_h^1}{p_h^0}\right)\theta\right) \times D^0 \tag{5.6}$$

5.3 CO_2 Conservation Supply Curve

After converting the energy savings into CO_2 abatement from previous work by Li/Zhu (2014), according to the energy savings of all BATs and the emission factor of each energy, we convert the energy savings of all BATs into CO_2 emissions. Figure 5.2 shows the CO_2 conservation supply curve when the discount rate is 20 %. A 20 % discount rate was selected mainly because it could reflect the expectation of payback period of steel enterprises (in general, the enterprises would like to have a short payback time and high internal rates of return when adopting such technologies). The CO_2 emissions per unit of steel for China's iron and steel industry are 1547.65 kg/ton steel output, and these 41 CATs could achieve 443.21 kg reductions in CO_2 emission, which would account for 22.3 % of total CO_2 emissions per unit of steel. Also, with an assumed CO_2 price which is equal to 0.1 yuan/kg, there are 28 cost-effective technologies. If we double the CO_2 price, the number of cost-effective CATs changes only slightly. Based on current allowance prices in China's pilot emission trading markets, the CO_2 price fluctuated from 0.05 yuan/kg (50 yuan/tCO_2) to 0.2 yuan/kg (200 yuan/tCO_2). Even with the CO_2 price as 200 yuan/tCO_2, it would still be little influence on the cost-effectiveness of the CATs. This is because a great number of CATs can generate negative adoption cost with the accounting of energy-saving benefits, so

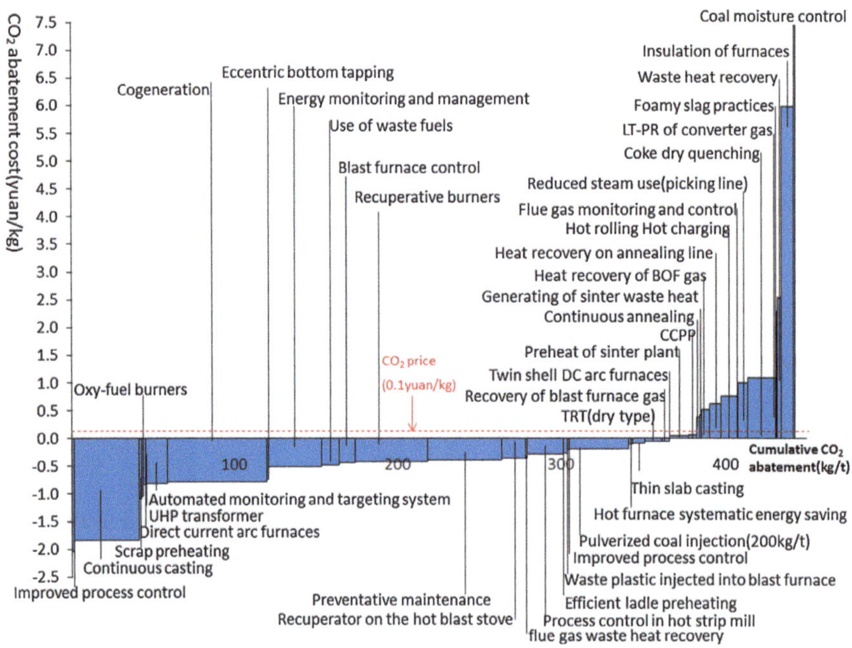

Fig. 5.2 CO_2 conservation supply curves of the discount rate 20 %. *Source* Li/Zhu (2014)

their CO_2 abatement costs are relatively low. Through our analysis, we found that only CATs that have positive costs will be influenced by the CO_2 price, which means, even with energy-saving benefit, some extra payment or loss will be made by enterprises in their adoption of these CATs (the number of such CATs is 16). Because there is no unique carbon market in China, we could not obtain a precise carbon price.

We now come to the CO_2 abatement cost curves of the specified process. Figure 5.3 shows the CO_2 abatement cost curves of sintering, coking, iron-making and BOF. The CATs of sintering and iron-making are all cost-effective. The CATs of the iron-making process have the largest cumulative CO_2 abatement. The CATs of coking and BOF are not cost-effective, and the cumulative CO_2 abatement of these two processes is also less than that of the iron-making process.

Figure 5.4 shows the CO_2 abatement cost of the EAF process. The CATs of the EAF process almost have good cost-effectiveness except for flue gas monitoring and control and foamy slag practices. Considering the small share of EAF (10 % in 2010), the cumulative CO_2 abatement is not very large. However, as the EAF share increases, the CO_2 abatement potential of the EAF process will become huge.

Figure 5.5 shows the CO_2 abatement costs of three additional processes. The CATs of the general technologies are all cost-effective, and also, this process has

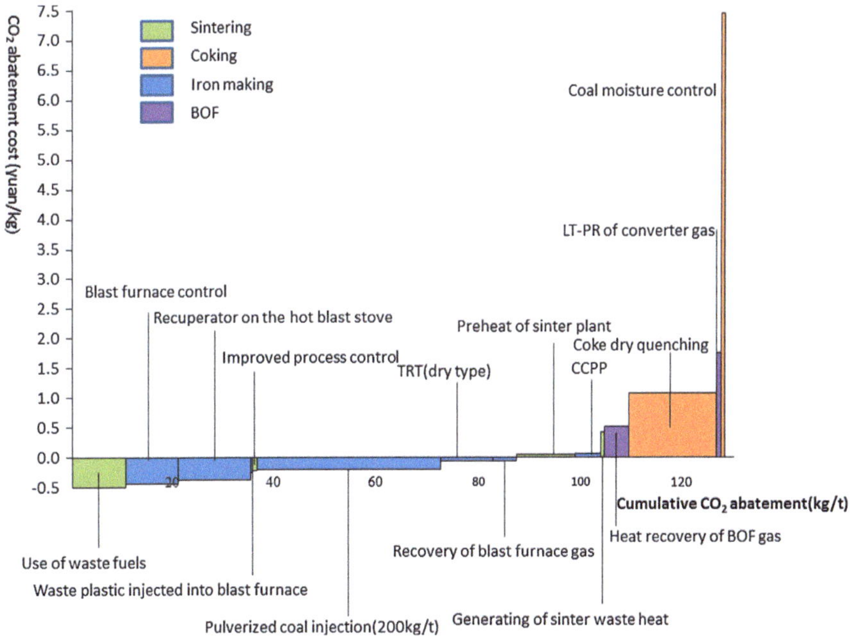

Fig. 5.3 CO_2 conservation supply curve of sintering, coking, iron making and BOF. *Source* Author's calculation

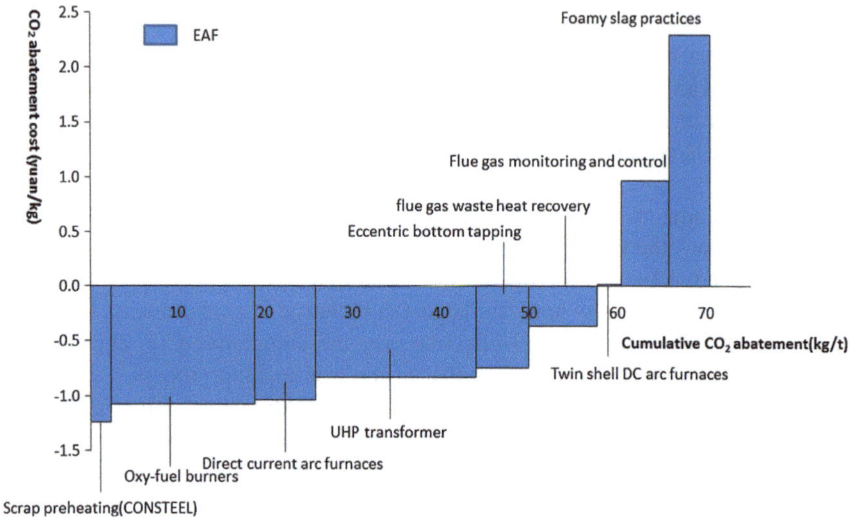

Fig. 5.4 CO_2 conservation supply curve of EAF. *Source* Author's calculation

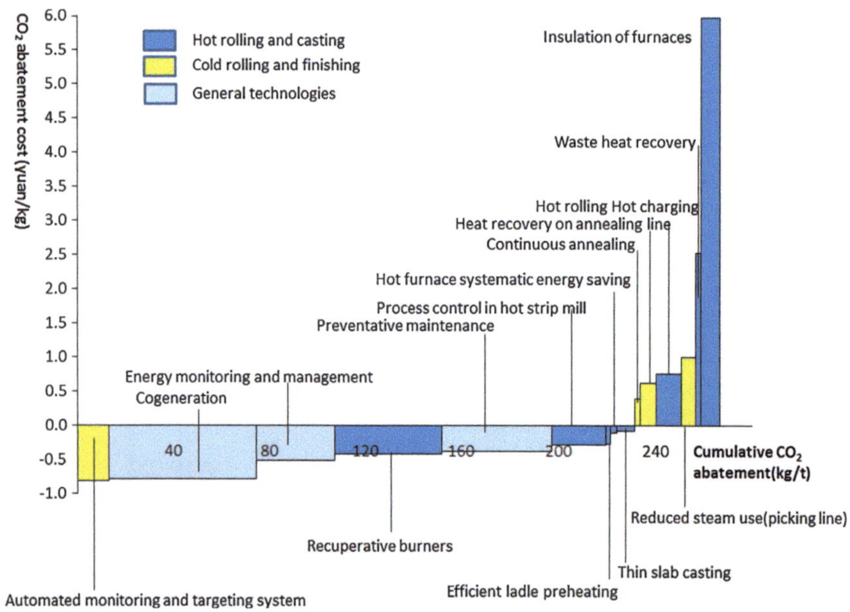

Fig. 5.5 CO_2 conservation supply curve of hot rolling and casting, cold rolling and finishing, general technologies. *Source* Author's calculation

the largest cumulative CO_2 abatement in this group. This is mainly because the CATs of this process are mutually beneficial and all have a larger adoption rate. The CATs of the cold rolling and finishing process are not cost-effective, except the automated monitoring and targeting system. Also, in the hot rolling and casting process, only half the CATs are cost-effective, and waste heat recovery and insulation of furnaces have very high CO_2 abatement costs, which will make them difficult to promote in the near future.

We should note that although the general technological process, casting and hot rolling currently have great CO_2 abatement, their adoption rate is almost maximised, so their future development potential is limited. EAF technologies are cost-effective, but their adoption rate is small. As the diffusion rate increases in the future, they will reach greater CO_2 abatement potentials.

5.4 CO_2 Abatement Potentials Analysis in 2020 and 2030 Under Different Situations

We have analysed the theoretical CO_2 abatement potential of these CATs as Fig. 5.6 shows. During this process, because about 1/3 of the crude steel of the world is produced by EAF, we assumed the share of BOF and EAF to be 7:3, which is approaching the advanced level of the world. We also assumed the CATs could reach a 100 % adoption. In this situation, the additional CO_2 abatement could

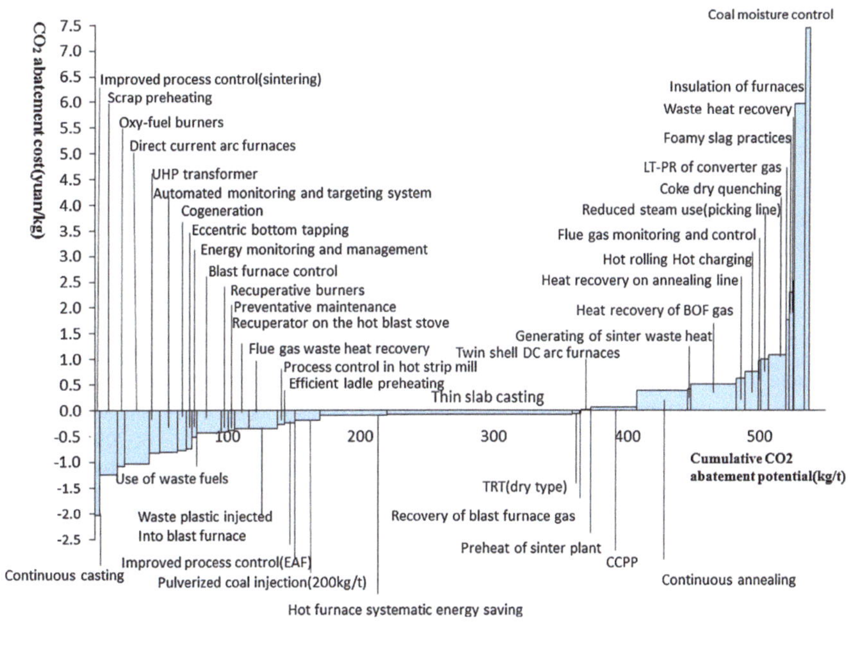

Fig. 5.6 Theoretical CO_2 abatement potential curve of the discount rate 20 %. *Source* Author's calculation

achieve 537.54 kg/t, which is more than the 2010 level. This is to say, the CO_2 abatement potential of China's iron and steel industry is still very large. To achieve this level, the EAF share should reach 30 %, and all CATs should be 100 % adopted —but since this situation is based on assumption, this abatement potential is also overestimated. Further, we found that the potentials of CATs with abatement costs below 0 to be very large (more than 300 kg/t)—these technologies both have cost-effectiveness as well as the abatement effect, and they should be promoted in the industry. We also noted that the CATs which had a higher adoption in 2010 tended to have lower CO_2 abatement potential due to the limitations for improvement.

We next forecasted the CO_2 abatement potentials of China's iron and steel industry in 2020 and 2030, respectively. Here we also referred to Li/Zhu (2014), and assumed three scenarios based on changing the diffusion rate, business as usual (BAU), cost-effective, and technical diffusion. The technical diffusion rate is exogenous. Because it is difficult to obtain data regarding the unique annual growth rate of each technology, we designed a share change per ten years to predict the change in the diffusion rate change. We gave unified share increases in 2020 and 2030, and these share increases were compared with those in 2010. The specific results are listed in Table 5.1.

The production of EAF is still very low in China (less than 10 % in 2010), mainly due to the scarcity of scrap steel and electricity. However, in recent years, the Chinese government has become more and more receptive to the advantages of EAF, such as the low level of investment and the ability to smelt special steel. It still has large development potential, so we assumed than by 2020 the share of EAF will have reached 20 % and that by 2030, it will hit 30 %. This percentage is close to the EAF levels of developed countries.

The CO_2 conservation supply curves of the three scenarios in 2020 are shown in Fig. 5.7. Figure 5.7 shows the differences between the three scenarios for 2020 and 2030. The higher the technical diffusion rate, the flatter the CO_2 abatement curve, and more emission abatement could be achieved by cost-effective CATs. The CO_2 abatement potentials also show an increasing trend over the same time period.

Table 5.1 Diffusion rate change under different scenarios for 2020 and 2030

Scenarios	Cost-effective technologies		Non cost-effective technologies	
	Share increase of 2020 compared with 2010 (%)	Share increase of 2030 compared with 2010 (%)	Share increase of 2020 compared with 2010 (%)	Share increase of 2030 compared with 2010 (%)
BAU	10	20	10	20
Cost-effective	20	40	10	20
Technical diffusion	20	40	20	40

Source Li/Zhu (2014)

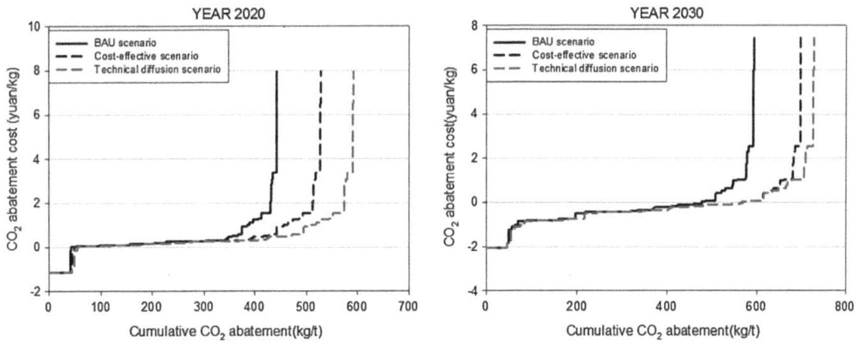

Fig. 5.7 CO_2 abatement curves of three scenarios in 2020 and 2030. *Source* Author's calculation

5.5 Impact of the Carbon Market on the Iron and Steel Industry

Because the carbon market will increase the production cost of enterprises, the loss of competiveness is a major concern for enterprises within the iron and steel industry. Since there is no clear definition of "competiveness loss", we think of it along the following lines: domestic production loss and domestic profit loss.

Here our MAC curve can be defined as shown in Formula (5.7):

$$\text{MAC} = \alpha \times ua + \beta \times ua^2 \tag{5.7}$$

In Formula (5.7), *ua* refers to the unitary abatement, and this MAC curve is fitted based on our CSC curve. Different allocation schemes will result in different sorts of anticipation. Because companies generally seek to maximise profits, and allocations do not take place every year, but every three or five years (in the European Union Emission Trading Scheme, EU ETS, they were first allocated every three years, then every five), the behaviour of these firms will depend on the allocation arrangement of the next period. We chose two allocation schemes—grandfathering and emission-based allocation—which are two main allowance allocation methods. Grandfathering is allocated according to the historical average emissions in the base period while emission-based allocation is based on the emissions in the last period.

5.5.1 Case 1: Grandfathering (GF) with Free Allocation (FA)

First, we defined Case 1, in which the allocation method is grandfathering and the share of FA is 50 %, meaning 50 % of the total carbon quota would be allocated to the enterprises freely. The purpose of free allocation is to maintain a reasonable level of profit, and to avoid dramatic profit losses.

We assumed that the CO_2 prices were 50, 80, 100 and 200 yuan/t, and obtained the following results. Unitary abatement's proportion of the total emissions is shown in Fig. 5.8. The proportion of unitary abatement increased as the CO_2 price increased; the higher the CO_2 price, the better the emission reduction effect. When the CO_2 price reached 200 yuan/t, the share of emissions reduction was over 50 %. The change of total emission reductions is shown in Fig. 5.9. Via a comparison with Fig. 5.8, we can see that the total emissions decreasing trend is similar to the trend for unitary abatement change. This means that the decrease in total emissions is mainly due to the increase in unitary abatement, and that the production change only impacts the total emissions change slightly.

The impact of various CO_2 prices on the iron and steel price, domestic production, total income and domestic profit compared with the 2010 baseline data is shown in Table 5.2. Based on the model of Chap. 2, when the CO_2 price fluctuated, the price, production, cost, income and profit changed with the CO_2 price. And when the CO_2 price changed, the impact of these changes on the key factors is

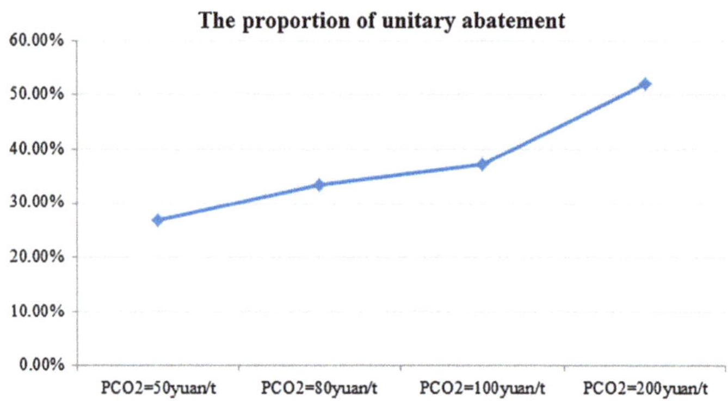

Fig. 5.8 The proportion of unitary abatement in iron and steel industry. *Source* Author's calculation

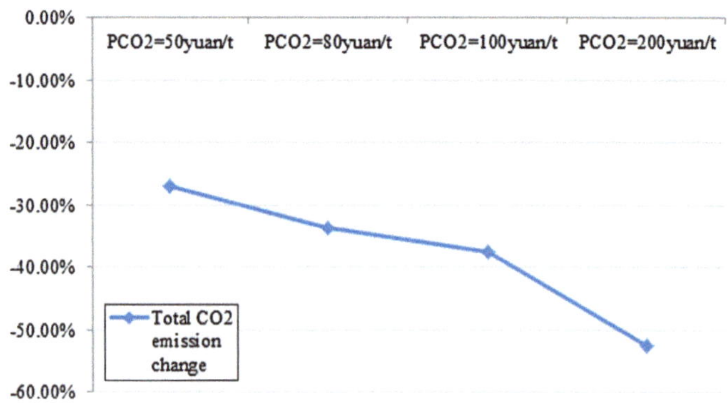

Fig. 5.9 The change of total emissions in Iron and Steel industry. *Source* Author's calculation

Table 5.2 The impact of different CO_2 prices on the key factors

	P_{CO2} = 50 yuan/t	P_{CO2} = 80 yuan/t	P_{CO2} = 100 yuan/t	P_{CO2} = 200 yuan/t
P_h (%)	0.62	0.85	0.96	1.15
q_h (%)	−0.18	−0.25	−0.29	−0.34
Total production cost (%)	1.11	1.64	1.95	3.11
Income change (%)	1.57	2.16	2.45	2.93
Profit change (%)	0.22	0.20	0.14	−0.58

Source Author's calculation

illustrated in Table 5.2. The higher carbon price would increase the price of domestic iron and steel products; and the higher the CO_2 price, the higher the domestic price. As a result, the domestic production will decrease due to the price increase; however, the CO_2 price will not greatly impact production. Total production went up as the CO_2 price increased, as did the income of enterprises. The profit change is very slight in this case. Only when the CO_2 price reached 200 yuan/t did the profit go down slightly, suggesting that the carbon market will not result in a huge loss of profits for the iron and steel industry.

In Case 1, the share of free allocation could help to maintain a reasonable profit level for enterprises. We also analysed as well how the free allocation share will affect profit change (Table 5.3), and found that the profit increased as the free allocation share increased. When FA was between 50 and 75 %, the enterprises' profit did not change much. The higher the percentage of free allocation was, the more the profit increased. However, considering the emission reduction target, the free allocation percentage should not necessarily be maximised.

The net purchase of sale allowance for the iron and steel industry is shown in Table 5.4. When the CO_2 price varied between 50–100 yuan/t, the iron and steel industry was a net buyer of CO_2 allocation. However, when the CO_2 price reached 200 yuan/t, the iron and steel industry then became a net seller in the carbon market.

Table 5.3 How the free allocation share affects the profit change

	P_{CO2} = 50 yuan/t (%)	P_{CO2} = 80 yuan/t (%)	P_{CO2} = 100 yuan/t (%)	P_{CO2} = 150 yuan/t (%)	P_{CO2} = 200 yuan/t (%)
FA = 100 %	0.11	0.22	0.96	0.54	0.82
FA = 90 %	0.05	0.12	0.58	0.37	0.59
FA = 75 %	−0.04	−0.02	0.01	0.10	0.24
FA = 50 %	−0.19	−0.26	−0.92	−0.34	−0.35
FA = 30 %	−0.31	−0.46	−1.66	−0.70	−0.82
FA = 0 %	−0.50	−0.75	−2.76	−1.24	−1.54

Source Author's calculation

Table 5.4 The net purchase of sale allowance of iron and steel industry

CO_2 price (yuan/t)	Net purchase (+) of sale (−) allowance
50	243.36
80	172.62
100	132.78
200	−23.32

Source Author's calculation. Unit: Mt

5.5.2 Case 2: Emission-Based (EB) Allocation

We then changed the allocation method, basing it on the emissions during the previous period. The free allocation of enterprise during period t can be defined as in Formula (5.8):

$$FA^t = q_e^{t-1} \times u_e^{t-1} \times s \quad (5.8)$$

Above, s refers to the stringency of the carbon allocation, and it could be adjusted by the government.

Here, we assumed three situations: s = 60 %, which means high stringency; s = 80 %, which means moderate stringency; and s = 90 %, which means low stringency. The impact of allocation stringency on key factors is shown in Table 5.5 (P_{CO2} = 50 yuan/t). The allocation stringency has an impact on domestic price, domestic production, exports and imports, export and import prices. When s = 90 %, the domestic price decreased, as did profits. This implied that the allocation stringency should be held around 80 %, which could keep profits stable.

Table 5.6 shows the impact of allocation stringency on profit, as well as on production. It shows that a lower allocation stringency results in higher profits than in Case 1. However, the profit on production is lower than Case 1. Considering the carbon market, the lower the allocation stringency, the higher the profit—because

Table 5.5 Impact of allocation stringency to key factors

	s = 60 %	s = 80 %	s = 90 %
p_h (%)	0.42	0.03	−0.16
q_h (%)	−0.13	−0.04	0.05
Total production cost (%)	0.04	0.12	0.21
Income change (%)	0.30	0.10	−0.11
p_m (%)	−0.49	−0.10	0.18
p_x (%)	0.31	0.10	−0.12
q_m (%)	−0.50	−0.10	0.19
q_x (%)	−0.27	−0.06	0.10
D (%)	−0.13	−0.04	0.05

Source Author's calculation

Table 5.6 The impact of allocation stringency to profit on production and profit

	s = 60 %	s = 80 %	s = 90 %	GF allocation (FA = 50 %)
Profit on production (%)	0.11	−0.07	−0.27	0.25
Profit (%)	−0.14	0.06	0.05	−0.19

Source Author's calculation

Table 5.7 Total emissions under Case 1 and Case 2 (P_{CO2} = 50 yuan/t)

Situation	Total emissions (Mt)	Compared with 2010 data (%)
s = 60 %	824.01	−26.93
s = 80 %	824.89	−26.85
s = 90 %	825.78	−26.77
Case 1	823.54	−26.97

Source Author's calculation

enterprises profit through the sale of allocations. If we consider the profit on production, a lower allocation stringency would decrease the profit that occurred during the production process. Actually neither of these two cases has much effect on industry profit (the change ranges between −0.3 and 0.3 %); in the EB case, the allocation stringency of 80 % could help to keep the industry profit and the profit on production stable (the change in both ranges between −0.1 and 0.1 %).

The total emissions under various allocation stringencies are shown in Table 5.7. The total emissions do not change greatly, because when the CO_2 price is fixed, the unitary emission is also fixed. Thus, as the production difference among these three situations is slight, the total emission difference is also slight. Compared with the 2010 data, total emissions decreased by about 27 %.

The net purchase or sell allocation under various allocation stringencies is shown in Table 5.8. As the allocation stringency increases, the iron and steel enterprises become net buyers in the carbon market; and when stringency is lower, the enterprises become net sellers in the carbon market.

Table 5.8 Net purchase or sell allocation under different situations

Situation	Net purchase (+) or sell allocation (−)
s = 60 %	138.19
s = 80 %	−14.80
s = 90 %	−176.46
Case 1	243.36

Source Author's calculation Unit: Mt

5.6 Conclusion and Policy Implications

We first studied the cost of CO_2 emission reductions in China's iron and steel sector. Forty-one CATs were selected based on various iron and steel production processes, and their investments, operation costs, CO_2 abatements and current shares in China's iron and steel industry were determined on the basis of references published in China. The CO_2 conservation supply curves for China's iron and steel industry were all calculated according to the data we collected.

We found that currently, general technologies, casting and hot rolling, and blast furnace have been making the largest CO_2 abatement contributions. This is probably the case since technologies involved in these processes are widely used. At the same time, it also means that their potential for growth is limited in the future. EAF technologies are almost all cost-effective. However, due to their low adoption rate, they have modest CO_2 abatement contributions at the present stage. Thus, we can predict that the share of EAF will increase in the future, and that the CO_2 abatement potential in the iron and steel industry will mainly come from EAF technologies. The development of EAF is mainly limited by the shortage of electricity and scrap, so the government should try to guide investment to improve the share of EAF and resolve technology barriers.

We then analysed the impact of the carbon market on China's iron and steel industry using partial equilibrium analysis. We defined Case 1 (a free allocation share is 50 %) and analysed the impact of an exogenous carbon price on the key factors in China's iron and steel industry, including domestic price, domestic production, imports and exports, import and export prices.

We established that a carbon market would increase the abatement cost and then increase the domestic price. The higher the CO_2 price, the higher will be the increase in domestic price, and as the result, domestic production will decrease—though only very slightly. In this case, the income of enterprises increased, and the change in profit was very little. Only when the CO_2 price reached to 200 yuan/t did the profit decrease slightly, which is to say that the CO_2 price did not result in a huge profit loss for the iron and steel industry. A carbon market could reduce total CO_2 emissions. However, most of the contribution would come from unitary emission reductions, not production reductions. When a carbon market increases import and export prices, imports will increase, but exports will decrease. At the same time, total consumption will also dwindle due to the rise in abatement costs.

Free allocation's largest contribution is how it helps maintain a reasonable profit level for the iron and steel industry. If the free allocation is zero, the profit of the enterprises will be decreased, no matter the price of CO_2. We found that the reasonable range for free allocation is about 50–75 %, which could ensure that the profit of enterprises does not fluctuate too greatly. With increasing allocation stringency, iron and steel enterprises will become net buyers in the carbon market, and when stringency is lowered, the enterprises will in turn become net sellers in the market.

Acknowledgments The authors acknowledge the financial support of the Chinese Academy of Sciences (No. XDA05150700) and the National Natural Science Foundation of China (No.71273253 and No.71210005). We deeply appreciate the weekly seminars at CEEP in CAS, from which improvements on an earlier draft of this paper were made.

References

Demailly, D. and P. Quirion (2008) "European Emission Trading Scheme and Competitiveness: A Case Study on the Iron and Steel Industry". *Energy Economics* 30: 2009–27.

Fan, J., Y. Wang and J.W. Liang (2004) "Study on the Import and Export of China". *Modern Economic Science* 26: 87–92.

Fischer, C. and A.K. Fox (2012) "Comparing Policies to Combat Emissions Leakage: Border Carbon Adjustments versus Rebates". *Journal of Environmental Economics and Management* 64: 199–216.

Fleiter, T., D. Fehrenbach, E. Worrell and W. Eichhammer (2012) "Energy Efficiency in the German Pulp and Paper Industry—A Model-Based Assessment of Saving Potentials". *Energy* 40: 84–99.

Guo, Z.C. and Z.X. Fu (2010) "Current Situation of Energy Consumption and Measures Taken for Energy Saving in the Iron and Steel Industry in China". *Energy* 35: 4356–60.

Hasanbeigi, A., C. Menke and A. Therdyothin (2010) "The Use of Conservation Supply Curves in Energy Policy and Economic Analysis: The Case Study of Thai Cement Industry". *Energy Policy* 38: 392–405.

Hasanbeigi, A., W. Morrow, J. Sathaye et al. (2012) *Assessment of Energy Efficiency Improvement and CO_2 Emission Reduction Potentials in the Iron and Steel Industry in China*. Berkeley, CA: Lawrence Berkeley National Laboratory.

Li, Y. and L. Zhu (2014) "Cost of Energy Saving and CO2 Emissions Reduction in China's Iron and Steel Industry". *Applied Energy* 130: 603–16.

Ministry of Industry and Information Technology (2010) "Total Energy Consumption Per Unit of Industry Added Value Decreased More than 25 % during the 11th Five-Year-Plan". http://finance.ifeng.com/news/20101124/2946133.shtml.

Price, L., A. Hasanbeigi, N. Aden et al. (2012) *A Comparison of Iron and Steel Production Energy Intensity in China and the U.S.* Berkeley, CA: Lawrence Berkeley National Laboratory.

Wang, K., C. Wang, X. Lu and J. Chen (2007) "Scenario Analysis on CO_2 Emissions Reduction Potential in China's Iron and Steel Industry". *Energy Policy* 35: 2320–35.

Worrell, E., N. Martin and L. Price (2000) "Potentials for Energy Efficiency Improvement in the US Cement Industry". *Energy* 25: 1189–214.

Zhang, B., Z. Wang, J. Yin and L. Su (2012) "CO_2 Emission Reduction within Chinese Iron & Steel Industry: Practices, Determinants and Performance". *Journal of Cleaner Production* 33: 167–78

Chapter 6
Energy Efficiency and Energy Conservation Strategies for Vehicles and Transport Systems in China

Hong Huo

Abstract Worldwide, the transport sector is a leading contributor to both energy use and greenhouse gas (GHG) emissions. It currently accounts for 19 % of global energy use and 23 % of global energy-related CO_2 emissions. In China, transport contributed 7–8 % to the nation's GHG emissions, but this value will increase rapidly in the coming years because China is in a period of significant growth in travel demand and vehicle ownership. Numerous actions have been taken to reduce the energy demand and GHG emissions for the transport sector, such as implementing fuel economy standards, promoting advanced vehicles and alternative fuels, and developing a high-speed railway system. This chapter analyses previous and soon-to-be implemented policies to control energy use and GHG emissions from transport in China, and then discusses the challenges and opportunities for China in building a low-carbon, clean, and sustainable transport system. Currently, technical measures to improve the fuel efficiency of transport dominate Chinese policy solutions, but a long-term strategy to decrease energy use and GHG emissions from China's transport system will likely rely on improving the railway transport system.

Keywords Energy efficiency · Greenhouse gases · New-energy vehicles · China

6.1 Introduction

The transport sector is one of the largest users of energy and emitters of greenhouse gases (GHG) in the world. According to estimates by the International Energy Agency (IEA), the transport sector represented 19 % of global energy use and 23 % of global energy-related CO_2 emissions (IEA 2008, 2009). These shares are even higher in developed countries. Figure 6.1 illustrates the contribution of transport to national energy use and GHG or CO_2 emissions in recent years in the US, Europe,

Dr. Hong Huo, Associate Professor, Institute of Energy, Environment, and Economy, Tsinghua University; Email: hhuo@tsinghua.edu.cn.

B. Su and E. Thomson (eds.), *China's Energy Efficiency and Conservation*,
SpringerBriefs in Environment, Security, Development and Peace 30,
DOI 10.1007/978-981-10-0737-8_6

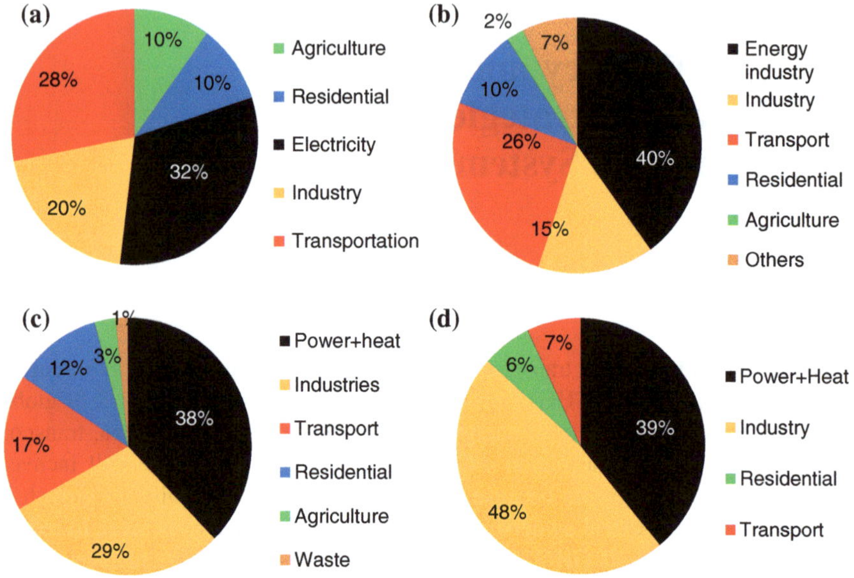

Fig. 6.1 Contribution by sector to the total GHG/CO_2 emissions in selected countries. *Source* US Environmental Protection Agency (2014), European Commission (2014), Ministry of the Environment of Japan (2014), MEIC (2014). **a** The US GHG (2012), **b** EU GHG (2012), **c** Japan GHG (2012), **d** China CO_2 (2010)

Japan, and China. As shown, transportation accounted for 28, 24, and 17 % of the GHG emissions in the US, Europe, and Japan, respectively. Figure 6.1d presents the contribution of the transport sector to the national CO_2 emissions in China in 2010 as estimated by the Multi-resolution Emission Inventory for China (MEIC), which was developed by Tsinghua University of China. Here, transport was estimated to contribute 7–8 % to China's CO_2 emissions, considerably lower than in the other nations. The manufacturing industry, the largest CO_2 emitter in China, accounted for 48 % of national CO_2 emissions, with two-thirds of emissions from this sector originating from fuel combustion and the remainder from industrial processes. Power and heating plants contributed approximately 39 % to national CO_2 emissions. The emissions from both of these sectors were driven by emissions from coal combustion, which is China's major source of CO_2 emissions. Although currently relatively small, because the demand for travel is growing significantly in China, the transport sector—which mainly consumes petroleum-based fuels—will become an important source of GHG emissions in the near future.

Combining the official statistics for revenue transport (National Bureau of Statistics of China 2013) and previous studies for private transport (Huo/Wang 2012; Huo et al. 2012) allows an estimation of the growth in transport demand in China. Such analysis suggests that the demand for passenger transport in China increased from 1,792 billion passenger-kilometres (Bpkm) in 2000 to 6,629 Bpkm in 2012, representing an annual growth rate of 11.5 %. For comparison, the

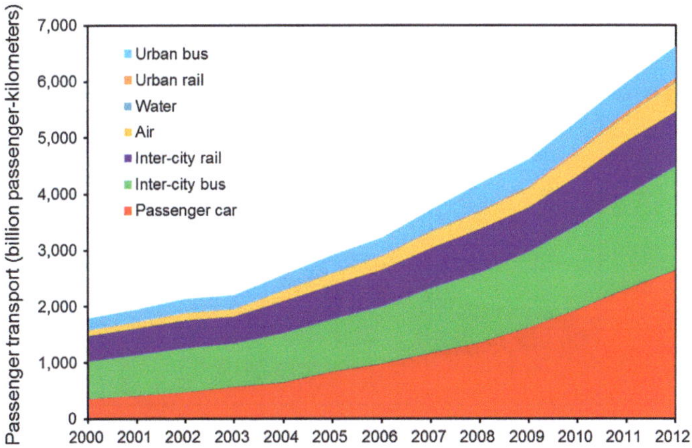

Fig. 6.2 Passenger transport (in passenger-kilometres) by mode in China, 2000–12. *Source* Results estimated using statistic data from National Bureau of Statistics of China (2013), and vehicle activity data from Huo et al. (2012)

corresponding data for 2000 and 2012 in the US were 8,186 and 7,789 Bpkm, respectively (US Department of Transportation 2014), while the figure grew from 5,372 to 6,391 Bpkm between 1995 and 2012 in the EU-28 countries (European Commission 2014).

Figure 6.2 presents the choice of different transport modes in China. Passenger-km travelled by passenger cars increased most rapidly during the period shown (from 20–40 % between 2000 and 2012) because of the tremendous growth in the number of private vehicles in recent years, while travel by inter-city buses also increased rapidly. This has been driven by a national effort to develop the national highway system and because—as a more flexible form of transport—more people have been choosing to travel by inter-city buses instead of train. Nevertheless, urban buses remain an important mode of transportation for commuting between home and work or school, accounting for 8–10 % of the total passenger-km travelled during the past decade. In total, road transport (passenger cars, inter-city buses, and urban buses) was responsible for 76 % of China's passenger transport. Similarly, inter-city trains, aeroplanes, and urban railways (subways) were responsible for 15, 8, and 1 %, respectively. Furthermore, the share of road transport will continue to grow in the near future, with most of this growth expected to come from passenger cars.

As Fig. 6.3 shows, the amount of freight transported into China also increased across the period, growing by a factor of three from 4,432 billion ton-kilometres (ton-km) in 2000 to 17,377 billion ton-km in 2012 (National Bureau of Statistics of China 2013). For comparison, between 2000 and 2012, freight transport in the EU-28 countries grew by less than 10 % (3,500–3,800 billion ton-km) (EU 2014) with even smaller changes observed in the US. In China, waterways accounted for the largest share of the freight transport sector (47 %), followed by road transport (34 %) and rail transport (17 %). In 2008, the Chinese Ministry of Transport began

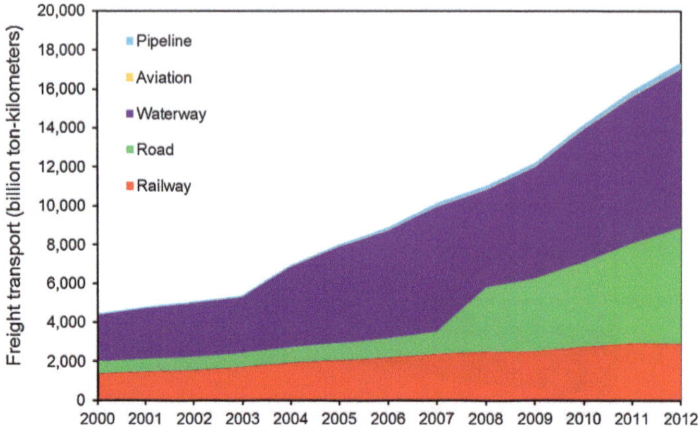

Fig. 6.3 Freight transport (in ton-km) by mode in China, 2000–12. *Source* National Bureau of Statistics of China (2013)

an investigation into the nation's road and waterway transport sectors. Through this study, a number of indicator statistics were modified, with these revisions being responsible for the significant change in data for these sectors for the period between 2007 and 2008.

Figure 6.4 presents the GHG emissions from the transportation sector in the US, EU-28, Japan, and China from 1990 to 2012. Transport-related GHG emissions in China remain less than those emitted by the US and EU-28. However, during the past decade, transport-related GHG emissions in the US, EU-28, and Japan have become constant and even begun to decrease while the transport-related GHG emissions in China have been increasing at an annual growth rate of 9 %.

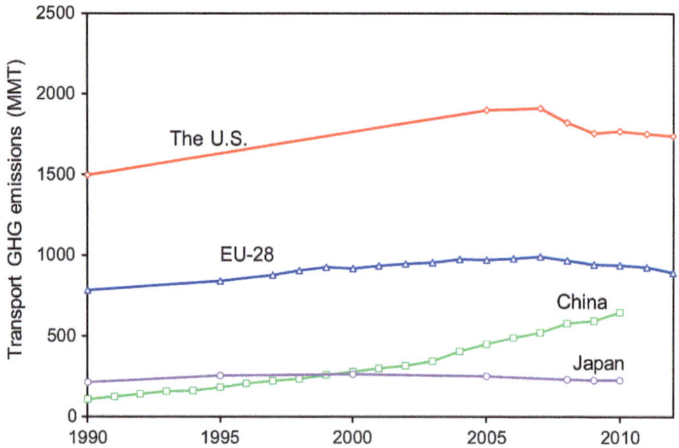

Fig. 6.4 GHG emissions from the transport sector from 1990–2012 in selected countries. *Source* US Environmental Protection Agency (2014), European Commission (2014), Ministry of the Environment of Japan (2014), MEIC (2014)

The importance of understanding the impact of China's transport sector and its associated policies on national energy security and the global climate has been emphasised in many previous studies (He et al. 2005; Yan/Crookes 2009; Huo et al. 2012a–d). This chapter analyses previous and soon-to-be implemented policies to control energy use and GHG emissions from transport in China and then discusses the challenges and opportunities for China in building a low-carbon, clean, and sustainable transport system.

6.2 Policies to Control Energy Use and GHG Emissions from Transport in China

6.2.1 Road Transport

Road transport accounts for approximately three quarters of the energy use of and GHG emissions from the global transport sector. Road transport is similarly the largest energy consumer and GHG emitter within the transport sector in most nations and therefore is necessarily involved in national strategies to limit these impacts. Within the numerous policies focusing on road transport that have been implemented around the world, three key factors are commonly targeted: vehicle stock; vehicle use intensity; and vehicle fuel efficiency. Effective methods of decreasing the energy use and GHG emissions from road transport should therefore be associated with decreasing vehicle use (the product of vehicle stock and vehicle use intensity) or increasing vehicle fuel efficiency. Other fuel-related options offer promise too, including using electric vehicles or alternative fuels that are based on energy sources other than petroleum. Most current Chinese policies in this area are focused on improving fuel efficiency and promoting electric vehicles and alternative fuels.

6.2.1.1 Fuel Economy Standards

Improving the fuel efficiency of vehicles is regarded as one of the most effective ways to reduce the energy use and associated GHG emissions from road transport and many countries continue to strengthen their requirements for vehicle fuel economy. Table 6.1 summarises the fuel economy standards implemented for each vehicle type in China.

In 2004, China issued its first vehicle fuel economy standards for light-duty passenger vehicles (GB 19578-2004). It was a two-phase standard that was expected to improve the fuel efficiency of passenger cars by 15 % by the end of the second phase. The standards were estimated to reduce the fuel consumption of light-duty vehicles from 9.11 L/100 km in 2002 to 7.80 L/100 km in 2009 (Wang et al. 2010; Huo et al. 2011). In 2012, the third phase (GB 27999-2011) was implemented. Starting in January 2015, this third phase requires fuel consumption

Table 6.1 Fuel economy standards implemented in China

Vehicle type	Standard No.	Issue date	Implementation date
Light-duty passenger vehicles			
Phase I	GB 19578-2004	2004	2005
Phase II	GB 19578-2004	2004	2008
Phase III	GB 27999-2011	2011	2012–15
Phase IV	GB 19578-2014; GB 27999-2014	2014	2016–20
Light-duty commercial vehicles	GB 20997-2007	2007	2008–09
Heavy-duty commercial vehicles	GB30510-2014	2014	2014–15

Source General Administration of Quality Supervision, Inspection and Quarantine of China, Standardization Administration of China (2004, 2011, 2007, 2014a, b, c)

rates of new passenger cars to be 6.9 L/100 km or less, which indicates a further 15 % improvement in their fuel efficiency. At the end of 2014, a fourth phase was introduced (GB 19578-2014), which limits new light-duty passenger vehicles to a fuel consumption rate of 5.0 L/100 km. A similar standard was issued in 2007 for light-duty commercial vehicles (e.g. light-duty trucks) (GB 20997-2007).

In 2010, the Ministry of Industry and Information Technology (MIIT) began reporting fuel consumption rates for light-duty vehicles (those of mass of 3,500 kg or less) and required each new light-duty vehicle to show the fuel consumption rate on a window label. These data are freely available, though they are based on results from a standard laboratory driving cycle that may underestimate real-world driving fuel consumption by 15.5 % (Huo et al. 2011).

Figure 6.5 presents historical and targeted fuel economy values in selected countries and shows that European countries have set a target of 5.5 L/100 km for passenger vehicles by 2015 (Wang et al. 2010; ICCT 2012). Although this is 20 % lower than the third phase of the fuel-economy standards in China (6.9 L/100 km by 2015), China's targets for passenger cars of 4.5–5 L/100 km by 2020 and 3.5–4 L/100 km by 2030 suggest the disparity between China and other countries could be eliminated within 10 years.

Japan was the first nation to issue a fuel economy standard for heavy-duty vehicles in 2006 (to be implemented in 2015). Meanwhile, the first US fuel economy standard for heavy-duty vehicles and buses was announced in 2011, targeting implementation in 2014–18. In China, the Ministry of Communications implemented industry standards to limit the fuel consumption rates of commercial vehicles for passenger transportation (JT 711-2008) and cargo transportation (JT 719-2008). Action in this area was further developed at the end of 2011 when MIIT released a national standard regulating the test methods for heavy-duty commercial vehicles (GB/T27840-2011) and a fuel consumption standard for heavy-duty industrial vehicles (QC/T924-2011). These standards were then further

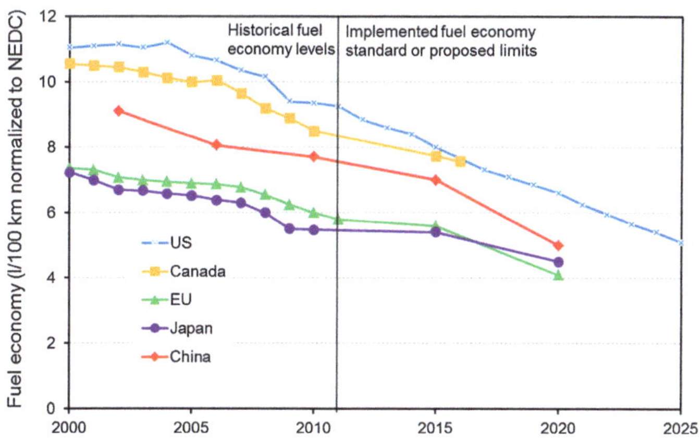

Fig. 6.5 Historical fuel economy levels and implemented or proposed fuel economy standards in selected countries. *Source* ICCT (2012)

updated in February 2014, with the issuance of Fuel Consumption Limits for Heavy-Duty Commercial Vehicles (GB30510-2014), which was implemented in July 2014.

6.2.1.2 Electric Vehicles

Most electric vehicles (EVs) are either pure battery electric vehicles or plug-in hybrid electric vehicles (HEVs). In China, EVs—or new-energy vehicles—are primarily understood to be pure battery electric vehicles. Global EV sales have increased significantly in recent years, from 100,000 in 2012 to approximately 300,000 in 2014. A similar situation has been observed in China, where EV sales grew from 10,000 in 2011 to over 50,000 in 2014, making it the second largest EV market in the world. The IEA's Electric Vehicles Initiative (EVI) involves China and 14 other countries and targets combined EV annual sales of 5.9 million by 2020 (IEA 2013).

EVs have been proposed in China as a potential solution to address the recent dramatic increase in petroleum-derived fuels for road transport. To this end, they have been promoted by a number of demonstration programs, such as the "Ten Cities and One Thousand Vehicles" program (Gong et al. 2012) and economic policies that favour the purchase of EVs. In April 2012, China's State Council approved the "Development Plan of Energy-Efficient and New-Energy Vehicles (2012–2020)", which plans to achieve accumulative sales of new-energy vehicles (including hybrids and EVs) numbering 500,000 by 2015, and 5 million by 2020 (China State Council 2012).

While EVs offer significant savings in terms of petroleum, the associated life-cycle CO_2 emissions need to be examined carefully. The CO_2 emissions from EVs depend on the CO_2 intensity of the electricity used to charge them (Huo et al. 2010), and in China the majority of electricity is generated from coal. In fact, in 2012 coal was responsible for 76 % of electricity generation with hydropower the second largest producer (16 %). Our recent analysis suggests that EVs may effect a 20 % decrease in GHG emissions compared with petroleum-fuelled vehicles and that, in regions where the share of coal-based electricity is relatively low, EVs may produce fewer emissions than even HEVs (Huo et al. 2013).

A combination of the decreasing carbon-intensity of China's electricity sector and developments in EV technologies suggests that the carbon intensity of EVs will also decrease. However, major obstacles to wider development of EVs include the construction of associated infrastructure and the development of battery technologies.

6.2.1.3 Alternative Fuels

Bio-fuels. In 2013, China accounted for approximately 3 % of global bio-ethanol production. This followed a 2001 decision to demonstrate ethanol-blended gasoline (E10, 10 % ethanol and 90 % gasoline by volume) for vehicles in selected locations in Henan, Heilongjiang, and Jilin provinces. The demonstration program was then expanded to all areas of six provinces (Heilongjiang, Jilin, Liaoning, Henan, Anhui, and Guangxi) and 27 cities in four other provinces (Hebei, Shandong, Jiangsu, and Hubei). The Government also established four grain-based ethanol production plants in Jilin (corn), Heilongjiang (corn), Henan (wheat), and Anhui (corn), and one cassava-based ethanol plant in Guangxi, with a 2014 total production capacity of approximately 2 million metric tons (MMT). The use of ethanol produced from food crops has been limited by the government and therefore only grown slowly. However, a 1 MMT ethanol plant based on cellulosic (non–food-derived) feedstocks is being built in Anhui marking significant steps towards a new bio-ethanol era in China. The 2007 "National Renewable Energy Plan for the Mid- and Long-term" issued by the National Development and Reform Commission suggests that the total consumption of fuel ethanol (based on grain or non-grain sources) could reach 10 MMT by 2020.

Alongside bio-ethanol production, China produced 0.8 MMT of bio-diesel in 2012, mostly from soybeans, jatropha, and used cooking oil. However, limited feedstock availability is expected to constrain the development of bio-diesel with the 'National Renewable Energy Plan for the Mid- and Long-Term' projecting consumption levels up to 2 MMT by 2020.

Similar to EVs, bio-fuels offer potential decreases in the use of petroleum, but their effects on energy use and GHG emissions need to be examined carefully from a life-cycle perspective. Per kilometre driven, corn-based ethanol has a limited benefit in reducing GHG emissions compared to gasoline if coal is used as the process fuel in the ethanol plant. However, GHG emissions may decrease by more than 30 % if using natural gas in place of coal. Moreover, ethanol derived from cellulosic sources can decrease GHG emissions by 80 % through the use of biomass to power the production processes.

*Coal-based fuels.*The use of coal-based methanol for powering vehicles has been demonstrated in Shanxi Province since the mid-1990s and most filling stations there now provide M15 (methanol-blended gasoline with 15 % methanol and 85 % gasoline by volume). A number of other provinces, such as Shanghai, Shaanxi and Zhejiang, have also recently begun promoting methanol-blended gasoline while demonstration programs for buses fuelled with di-methyl ether or Fischer-Tropsch diesel are also ongoing.

Although coal-based fuels offer decreasing consumption of petroleum products, from a GHG perspective coal-based fuels were reported to double the life-cycle CO_2 emissions compared with conventional fuels, or equal them if carbon capture and sequestration technology is employed (Wang/Huo 2009). Coupled with the fact that increasing air pollution worries in China are likely to influence where and how coal is used in the future, the environmental impacts of coal use have the potential to affect the development of coal-based transportation fuels.

Compressed natural gas (*CNG*). Programs for CNG vehicles (CNGVs) have been running in China since the late 1990s, and by June 2013 the nation had approximately 3,000 CNG gas stations. As of May 2012, China had approximately 1.1 million CNGVs, accounting for approximately 60 % of all buses and taxis (International Association for Natural Gas Vehicles 2012). In some regions, such as Chongqing and Sichuan, the share of taxis and buses fuelled with CNG is more than 95 % and this growth trend is expected to continue.

The life-cycle GHG emissions of CNGVs vary according to the distance that the natural gas is transmitted. For regions located close to natural gas sources, CNGVs can exhibit GHG emissions that are 10 % lower than gasoline cars, while for regions located far from the source (e.g. Shanghai and Zhejiang), CNGVs may increase the life-cycle GHG emissions (Huo et al. 2013).

6.2.1.4 Reducing Vehicle Use

Some cities in China have restricted the sale and use of vehicles to address serious traffic congestion and urban air pollution problems—the details of these programs are summarised in Table 6.2. Shanghai was the first city to restrict vehicle sales, when in 1994 the city began a licence plate auction policy. The limited number of new licence plates available each year (e.g., 110,000 plates in 2013) has caused new licence plates to be very expensive: the average cost was 80,000 yuan (equivalent to US$12,500) in 2013. Beijing began limiting vehicle sales to 20,000 per month in 2011 (40 % of the average monthly sales in 2010) with sales quotas allocated to potential buyers through a lottery process. The quota was further tightened to approximately 12,000 vehicles in January 2014. In 2012, Guangzhou initiated a similar policy to restrict vehicle sales to 120,000 per year but also further specified the system and allocated 10 % of the sales to new energy vehicles (e.g. electric vehicles), 40 % to be determined by a lottery system, and 50 % by auction. Other

Table 6.2 City-level policies on vehicle sale restriction in China (as of March 2015)

City	Policy start date	Annual quota	Allocation pattern
Shanghai	1994	<120,000	Auction
Beijing	Jan, 2011	240,000	Lottery
	Jan, 2014	150,000	Lottery
Guiyang	July, 2011	24,000	Lottery
Guangzhou	July, 2012	120,000	Action (40 %) + lottery (50 %) + NEV (20 %)
Tianjin	Dec, 2013	100,000	Action (40 %) + lottery (50 %) + NEV (20 %)
Hangzhou	Mar, 2014	80,000	Action (20 %) + lottery (80 %)
Shenzhen	Dec, 2014	100,000	Action (40 %) + lottery (40 %) + NEV (20 %)

NEV refers to new energy vehicles. *Source* summarised by the author from news via the Internet

cities such as Shenzhen, Guiyang, and Tianjin have adopted similar restriction policies.

As well as focusing on new car sales, various policies have been implemented at the city-level to restrict private cars, yellow-labelled vehicles (pre-State I gasoline vehicles, and pre-State I, State I and State II diesel vehicles), trucks, or those vehicles registered in other provinces to enter into certain areas on certain days. In October 2008, Beijing introduced a policy to prevent 20 % of private cars entering urban areas during the daytime on workdays. The prevented portion of the fleet is determined according to the last digit of the licence plate number. Thus, for example, those with licence plate numbers ending in 1 or 6 are not allowed to drive in urban areas of Beijing on Mondays, and those ending in 2 or 7 are prevented from driving on Tuesdays. The restrictions are then rotated every 3 months. During special events, such as the 2008 Olympics Games and the 2014 OPEC meeting, the Beijing government restricted vehicle use by only allowing 50 % of cars to drive in the city.

While this policy has helped to alleviate traffic congestion and reduce localised vehicle energy use and emissions significantly, economic measures to restrict vehicle use—such as raising parking fees and charging congestion fees in urban centres—are also currently under discussion.

6.2.2 Non-road Transport

Non-road transport, which includes railways, aviation, and waterways, is also subject to targets aimed at decreasing fuel consumption. The 12th Five-Year Plan for the transport sector projected that fuel consumption per passenger-km or ton-km would decrease by 5 % between 2010 and 2015 for railway transport; by 15 % between 2005 and 2015 for waterway transport, and by 3 % between 2010 and

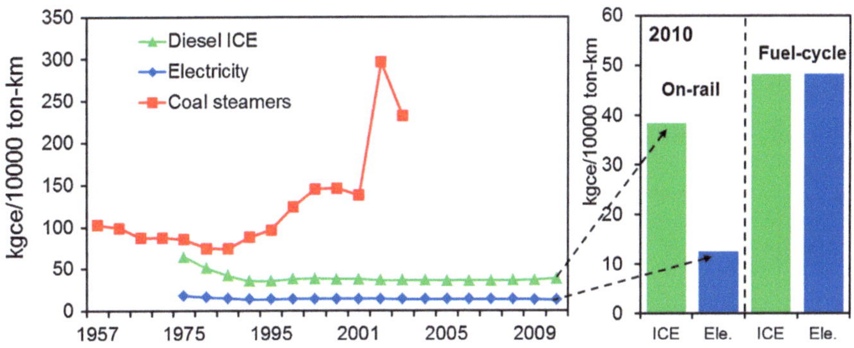

Fig. 6.6 Fuel efficiency of different locomotive engine technologies in China, 1957–2010. *Source* Railway Statistics Centre of China (2008–2012), fuel-cycle results are calculated by the author

2015 for aviation transport (Ministry of Transport of China 2011; National Railway Administration of China 2012).

Transport by railway is important in China's transport system and the sector has undergone significant changes in recent decades. While steamer engines—which were powered by coal and had poor energy efficiencies—were phased out in 2003, the electrified share of national passenger transport increased from 32 to 65 %. As Fig. 6.6 shows, although electric locomotives consume 50 % less energy than diesel per ton-km on the railway, because approximately 70 % of electricity is produced from coal in China, the two engines perform similarly from a life-cycle GHG-perspective.

Another significant change in China's transport sector is the rapid growth of high-speed trains. Between 2008 and 2012 alone, the share of passenger transport carried by high-speed rail increased from 0.2 to 14.7 %, approaching the current levels observed in European countries and in Japan (see Fig. 6.7a). By 2011, high-speed lines covered 9,400 km in China, significantly more than any other nation (Fig. 6.7b). While the energy consumption rates of high-speed trains are estimated to be higher than those of ordinary trains because of the need to overcome greater wind-resistance, some reports in Europe suggest that the greater occupancy rates compensate for this and that their energy use per passenger-km is lower than that of ordinary trains. China is planning to further expand the high-speed system by some 30,000–50,000 km. It is hoped that the convenience, comfort, and time-savings that high-speed trains offer will attract people and draw them away from the most energy-consuming transport modes (cars and aeroplanes) for medium-length trips (100–1,000 km). Were such a shift to occur, China's large-scale plans to develop high-speed trains could significantly help to decrease the transport sector's energy use and GHG emissions.

The energy efficiency of aircraft operating in China has increased in recent years, with fuel consumption per ton-km decreasing by 30 % between 2000 and 2012 (Fig. 6.8). Further enhancements from new technologies are unlikely to significantly affect this situation because most aircraft currently operating in China are

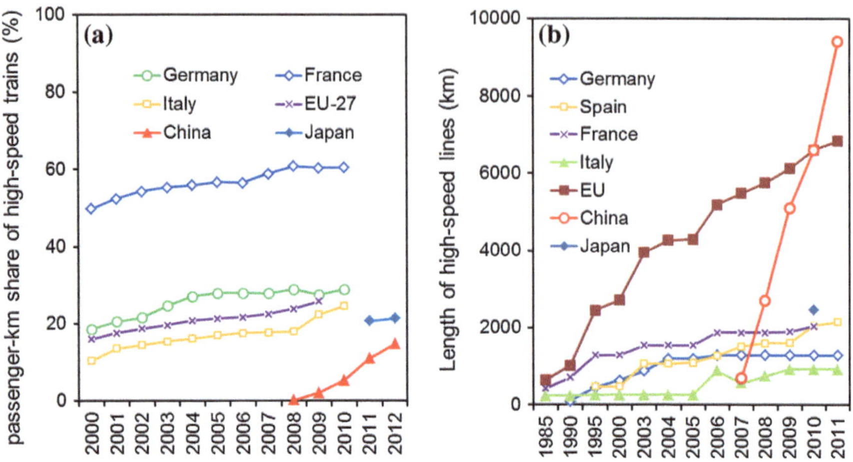

Fig. 6.7 Comparison of high-speed train systems in several countries. *Source* European Commission (2014), National Bureau of Statistics of China (2013), Statistics Bureau of Japan (2013)

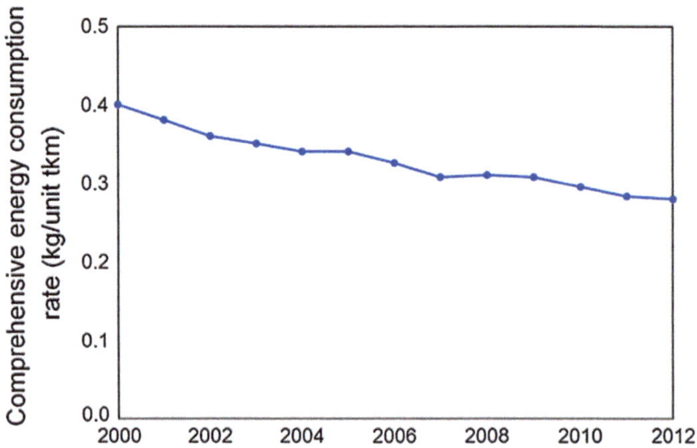

Fig. 6.8 Comprehensive energy consumption rates of civil aircraft in China (2000–12). *Source* Civil Aviation Administration of China (2000–2012)

similar to those in use elsewhere as they have been imported from developed countries. However, another factor that influences aircraft fuel consumption is the load factor, which has averaged approximately 80 % for passenger transport and 70 % for freight transport in recent years (Civil Aviation Administration of China 2000–2012). Thus, improving load factors could offer improvements in energy use and GHG emissions, and policies to achieve such increases should be included in the energy strategies of the civil aviation sector in the near future.

6.3 Challenges and Opportunities

Before 2000, the level of vehicle ownership in China was very low, with public modes of transport—such as urban buses, the subway, and railways—being the main ways used to get around. Now, China is in a rapid transition period with its population rapidly switching from public transport to private cars. This switch and the growth in demand for transport has caused serious traffic congestion and air pollution in urban areas as well as significant growth in the energy use of and GHG emissions from the transport sector.

The question of how to deal with the dramatic growth in travel demand and individuals' desire to own and drive cars remains a major challenge for China's transport sector. A failure to manage this challenge will inevitably see a continued increase in the energy use and GHG emissions of China's transport sector in the near future. Current policies to mitigate the impacts of the transport sector are focused mainly on promoting energy-saving technologies, but their positive effects could be offset or outweighed by the growth in travel demand and vehicle use.

Restrictions on the sale and use of vehicles can benefit cities at a local level but have limited impact nationally. However, longer-term measures such as developing large fleets of new energy vehicles and effective public mass transit systems could deliver great benefits in the future.

As well as technological steps, various economic measures should be considered to encourage the use of new energy vehicles and public mass transit, such as the offering of subsidies for purchasing new energy vehicles and discounts for monthly and annual public transit passes.

In 2014, China was presented with a significant opportunity to develop a low-carbon transport sector with the release of the US-China Joint Announcement on Climate Change, in which the two countries committed to cooperate intensively in developing EVs, the transport sector, and wider low-carbon technologies.

Addressing the air pollution issues that vehicles are currently causing will have some impact on the associated GHG emissions from the transport sector. However, the ultimate solution to decrease energy use and GHG emissions from vehicles will rely on the diversity in both the fuel and vehicle technologies, and the modes of travel used.

References

China's State Council (2012) *Development Plan of Energy-Efficient and New-Energy Vehicles (2012–2020)*. Beijing. http://www.gov.cn/zwgk/2012-07/09/content_2179032.htm [accessed 31 July 2012].

Civil Aviation Administration of China (2000–12) *China's Aviation Yearbook*. Beijing: China Civil Aviation Press.

European Commission (2014) *EU Energy in Figures: Statistical Pocketbook 2014*. Luxembourg: Publications Office of the European Union.

General Administration of Quality Supervision, Inspection and Quarantine of China, Standardization Administration of China (2004) *Limits of Fuel Consumption for Passenger Cars* (GB 19578-2004).

General Administration of Quality Supervision, Inspection and Quarantine of China, Standardization Administration of China (2007) *Limits of Fuel Consumption for Light Duty Commercial Vehicles* (GB 20997-2007).

General Administration of Quality Supervision, Inspection and Quarantine of China, Standardization Administration of China (2011) *Fuel Consumption Evaluation Methods and Targets for Passenger* Cars (GB 27999-2011).

General Administration of Quality Supervision, Inspection and Quarantine of China, Standardization Administration of China (2014a) *Fuel Consumption Limits for Passenger Cars* (GB 19578-2014).

General Administration of Quality Supervision, Inspection and Quarantine of China, Standardization Administration of China (2014b) *Fuel Consumption Evaluation Methods and Targets for Passenger Cars* (GB 27999-2014).

General Administration of Quality Supervision, Inspection and Quarantine of China, Standardization Administration of China (2014c) *Fuel Consumption Limits for Heavy-Duty Commercial Vehicles* (GB30510-2014).

Huo, H. and M. Wang (2012) "Modeling Future Vehicle Sales and Stock in China". *Energy Policy* 43: 17–29.

Huo, H., Q. Zhang, M. Wang et al. (2010) "Environmental Implication of Electric Vehicles in China". *Environ. Sci. Technol.* 44: 4856–61.

Huo, H., Z. Yao, K. He and X. Yu (2011) "Fuel Consumption Rates of Passenger Cars in China: Labels versus Real-World". *Energy Policy* 39: 7130–5.

Huo, H., Q. Zhang, K.B. He et al. (2012) "Vehicle-Use Intensity in China: Current Status and Future Trend". *Energy Policy* 43: 6–16.

Huo, H., Q. Zhang, F. Liu, and K. He (2013) "Climate and Environmental Effects of Electric Vehicles versus Compressed Natural Gas Vehicles in China: A Life-Cycle Analysis at Provincial Level". *Environ. Sci. Technol.* 47: 1711−8.

ICCT (International Council on Clean Transportation) (2012) *Global Comparison of Light-Duty Vehicle Fuel Economy/GHG Emissions Standards.* http://www.theicct.org/sites/default/files/info-tools/ICCT_PVStd_June2012_v0.pdf [accessed 29 May 2014].

IEA (International Energy Agency) (2009) *Transport, Energy and* CO_2*: Moving toward Sustainability.* https://www.iea.org/publications/freepublications/publication/transport2009.pdf [accessed 29 May 2014].

International Energy Agency (2013) *Global EV Outlook: Understanding the Electric Vehicle Landscape to 2020.* https://www.iea.org/publications/freepublications/publication/GlobalEV Outlook_2013.pdf [accessed 1 August 2013].

International Association for Natural Gas Vehicles (2012) *China's NGV Growth Accelerating.* http://www.iangv.org/2012/06/chinas-ngv-growth-accelerating/ [accessed 31 July 2012].

Ministry of the Environment, Japan (2014) *National Greenhouse Gas Inventory Report of Japan.* Ibaraki, Japan: National Institute for Environmental Studies.

Ministry of Transport of China (2011) *12th Five-Year Development Plan for Transport.* http://www.moc.gov.cn/zhuantizhuanlan/jiaotongguihua/shierwujiaotongyunshufazhanguihua/ [accessed 29 May 2014].

National Bureau of Statistics of China (2013) *China Statistical Yearbook 2013.* Beijing: China Statistics Press.

National Railway Administration of China (2012) *12th Five-Year Development Plan for Railway.* http://www.nra.gov.cn/fwyd/hyjc/ghfz/tlhygh/tlsewfzgh/ [accessed 29 May 2014].

Railway Statistics Centre of China (2008–2012) *Railway Statistics Abstract.* Beijing: China Railway Publishing House.

Statistics Bureau of Japan (2013) *Japan Statistical Yearbook 2013.* Tokyo: Statistics Bureau of Japan.

US Environmental Protection Agency (2014) "Inventory of U.S. Greenhouse Gas Emissions and Sinks: 1990–2012". US EPA report, EPA 430-R-14-003.

US Department of Transportation (2014) *U.S. Passenger-Miles*. http://www.rita.dot.gov/bts/sites/rita.dot.gov.bts/files/publications/national_transportation_statistics/index.html [accessed 31 July 2015].

Wang, Z., Y. Jin, M. Wang and W. Wu (2010) "New Fuel Consumption Standards for Chinese Passenger Vehicles and Their Effects on Reductions of Oil Use and CO_2 Emissions of the Chinese Passenger Vehicle Fleet". *Energy Policy* 38: 5242–50.

Chapter 7
Building Life-Cycle Commissioning and Optimisation: Approach and Practice

Shengwei Wang and Dian-ce Gao

Abstract The energy consumption of buildings has increased rapidly worldwide in recent years. Building systems usually do not work as efficiently as intended due to various faults in their life-cycle. This paper presents a series of life-cycle commissioning and optimisation approaches that intend to realise the energy efficient operation of the overall heating, ventilation and air conditioning (HVAC) system practically. At design stage, optimisations and commissioning are performed to optimise the system configuration and components selection, making the designs proper and correct. At the construction stage, commissioning is conducted to construct and install systems (e.g. HVAC components, instrumentations) correctly. At the testing and commission and operation stages, commissioning is made to ensure systems operate as well as the design intent in real operations. Optimal control strategies are developed and implemented to push the system approach to its best performance, often exceeding design intent (normal standard). A simplified energy performance assessment method based on macroscopic energy balance principles is developed for assessing the performance of information-poor buildings. The proposed methods have been implemented in many new and existing real buildings. Significant energy benefits have been achieved.

Keywords Building energy · Life-cycle · Commissioning · Design optimisation · Optimal control · System diagnosis · Energy assessment

7.1 Background of Building Energy

The energy consumption of buildings has increased rapidly in recent years due to population growth, the rising demand for healthy, comfortable and productive indoor environments, global climate change, etc. Globally, energy consumption

---

Dr. Shengwei Wang, Professor, Department of Building Services Engineering, The Hong Kong Polytechnic University; Email: beswwang@polyu.edu.hk.

Dr. Dian-ce Gao, Department of Building Services Engineering, The Hong Kong Polytechnic University.

B. Su and E. Thomson (eds.), *China's Energy Efficiency and Conservation*,
SpringerBriefs in Environment, Security, Development and Peace 30,
DOI 10.1007/978-981-10-0737-8_7

from buildings is approximately 40 % (Omer 2008). In the United States, buildings accounted for 74 % of electricity use in 2010 (DOE 2012); and in China, the building sector consumed nearly 20 % of China's total primary energy consumption in 2006, in which electricity comprised 44 %. In Hong Kong, the energy consumption of buildings in 2009 took up nearly 91 % of the total electric energy consumption.

Heating, ventilation and air conditioning (HVAC) systems occupy about 50 % of building energy use on average (Perez-Lombard et al. 2008). However, excessive energy is consumed in building HVAC systems because they often fail to operate as intended, after a period of time, even with correct commissioning. These failures may be caused by inappropriate design or improper operation—such as inappropriate temperature set points, poor outside air damper control, as well as sensor drifts, etc. Proper operation of building systems can lead to improved occupant comfort and health, greater energy efficiency, longer lifespan of equipment, reduced maintenance costs, reduced unscheduled equipment shutdown time, and so on.

7.2 Basic Approaches for Saving Energy in Buildings

Generally, energy saving in buildings can be achieved by the following three approaches.

7.2.1 Reduced Heating/Cooling Loads

The heating/cooling load refers to the required heating/cooling provided by the HVAC system to maintain the indoor thermal comfort at a predefined level. Reducing the heating/cooling load will accordingly reduce the energy consumed by HVAC systems.

The heating/cooling load mainly depends on factors such as weather conditions, building envelopes (e.g. walls and windows) and indoor heat sources (e.g. occupancy, lighting, equipment). To reduce heating/cooling load, one could optimise the building envelope; for instance, the use of thicker walls or walls with lower overall heat transfer coefficient would significantly lower the heat transfer between indoor and outdoor air.

The optimisation of the building envelope is usually referred to as *passive design*, which can be performed at the design stage. By modelling the building using professional software tools (e.g. Energy Plus), the building energy performance with different wall/window materials, orientations and shielding, etc. can be assessed. The building envelopes with optimal trade-offs between energy performance and initial cost will then be selected as the preferred option.

7.2.2 *Use of Energy Efficient Components and Technologies*

For the HVAC system, the use of energy efficient components and technologies is another way to achieve energy savings while satisfying the same heating/cooling demand.

Energy efficient components refer to the equipment of HVAC system, such as the chillers, pumps and fans, which hold higher energy efficiency than the traditional alternatives. For instance, a water-cooled chiller is more energy efficient than the air-cooled chiller.

Energy efficient technologies, here, refer to the advanced techniques that can achieve the same functions using less energy, such as liquid desiccant technology and ground source heat-pumps.

Obviously, the proper selection and use of these energy efficient components and technologies are supposed to achieve energy-saving benefits. This straightforward approach has also been widely implemented in practical applications.

7.2.3 *System Optimisation and Optimal Control*

An HVAC system comprises many sub-systems and components. The proper adoption of energy efficient components and technologies alone is insufficient. Without proper system configuration, good integration and proper control, the individual components put together will not work as efficiently as anticipated. Therefore, in order to ensure the enhanced energy performance of the overall system, it is essential to make good designs by optimising system configurations, and to realise energy efficient operations by proper commissioning and controls. Although system optimisation and optimal control have not yet attracted the necessary attention, some 20–50 % of energy savings can be expected from their implementation (Claridge et al. 1996, 2000; Liu et al. 1994).

7.3 Concept of Building Life-Cycle Commissioning and Optimisation

7.3.1 *Overview*

Life-cycle commissioning and optimisation is geared to realising the practical energy efficient operation of the overall HVAC system practically. It covers the design stage, construction stage, testing and commissioning stage, and operation stage in the life-cycle of an HVAC system, as shown in Fig. 7.1.

At the design stage, optimisations and commissioning are performed to enhance the system configuration and components selection, making the designs proper and

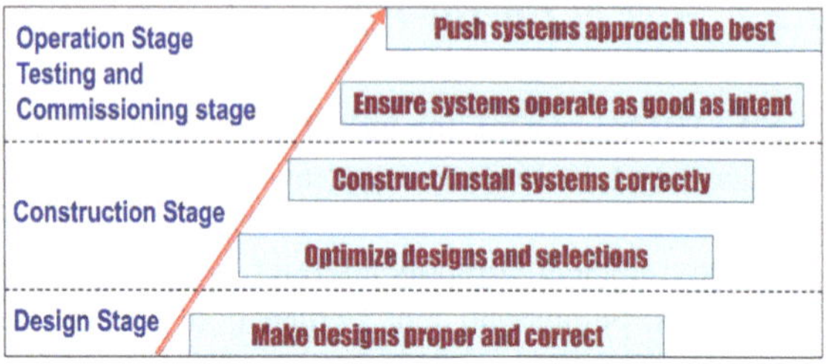

Fig. 7.1 Steps towards energy efficient buildings. *Source* Authors' work

correct. At the construction stage, commissioning is conducted to construct and install systems (e.g. HVAC components, instrumentations) correctly. At the testing and commission and operation stages, commissioning is carried out to ensure that systems operate in practice to meet the design intent. Optimal control strategies are developed and implemented to push the system towards achieving its best performance, and often to exceed design intent (normal standard).

7.3.2 Concept and Objective of Commissioning

In IEA (International Energy Agency) Annex 40, commissioning is defined as a "quality-oriented process for achieving, verifying and documenting whether the performance of a building's systems and assemblies meet defined objectives and criteria". It is quite close to the definition used by ASHRAE in which commissioning is referred to as "the process of ensuring that systems are designed, installed, functionally tested and capable of being operated and maintained to perform in conformity with the design intent".

Commissioning, here, includes verification, diagnosis, correction and improvement of the system design and operation. Its objective is to ensure that the operation performance of the systems meet the design intent from design stage to operation stage. Since the performance deviations would come from different stages of the building life-cycle (i.e. design, installation, testing and commissioning, and system operation/maintenance), diagnosis and verification should be conducted through the entire life-cycle.

As early as the design stage, diagnosis and verification can be performed to evaluate system configurations. For instance, Ma et al. (2008a) proposed an alternative design for a complex secondary chilled water system, in which a great amount of annual energy saving (1,000,000 kWh) was projected. Diagnosis and verification at this early stage also allows for greater flexibility in system configuration improvement since the system has not been installed yet.

Many diagnosis strategies have also been applied in the testing and commissioning stage, in which the building services system (e.g. lighting, HVAC) as well as their metering system are function-tested. Such diagnosis belongs to the offline fault detection and diagnosis (FDD) category and does not require integration with building automation systems (BAS). A strategy is developed to automatically diagnose and evaluate the sensors and building refrigeration systems during commissioning (Wang 2002).

At operation stage, diagnosis has great significance in ensuring the healthy operation of components and sensors to enhance the system's overall performance. Algorithms are usually used to identify the causes of the performance deviation of a system from its design intent (e.g. degraded components, failed sensors). Online FDD strategies are typically performed at the system operation stage. Different levels of FDD strategies—e.g. sensor level (Xiao et al. 2006), component level (Nassif et al. 2008) and system level (Wang et al. 2010)—have been developed and validated.

7.3.3 Concept and Objective of Optimisation

Optimisation is an area of mathematics that is concerned with searching for the best points, curves and surfaces, etc. (Hull 2003). The key is to find the global, not local, optimal solution.

Optimisation, here, aims to ensure HVAC systems provide expected quality of services (i.e. comfortable and healthy environments) with minimum energy consumption. According to the optimisation problems encountered at different stages in a building's whole life cycle, the optimisation of HVAC systems can be categorised into design optimisation and operation/control optimisation.

7.3.3.1 Design Optimisation

Different design configurations and various component selections may result in different initial costs and energy use. The design optimisation therefore aims at searching for optimal design configurations and most suitable components that ensure the systems consume the least energy consumption at operation with a reasonable initial cost. In practice, the design engineers are responsible for considering various systems and recommending one or more that meet the goal and perform as desired based on optimisations (ASHRAE 2008).

7.3.3.2 Operation/Control Optimisation

Operation/control optimisation aims at pushing the operational performance of HVAC systems towards achieving their best performance, and often exceeding the design intent, by using optimal control strategies.

Optimal control strategies are developed and implemented to instruct the HVAC system to operate with enhanced energy efficiency while still satisfying the indoor thermal comfort and healthy environment. The outputs of these optimal control strategies mainly include optimised sequencing, speed, set-points of temperature and pressure, valve openings, etc. For instance, Ma/Wang (2009a) presented optimal control strategies for variable speed pumps with different configurations in complex building air-conditioning systems to enhance their energy efficiencies, in which an optimal pump sequence control strategy was developed to determine the optimal number of pumps in operation.

7.4 Methodology and Approaches

7.4.1 Dynamic Virtual Test Platform for Commissioning and Optimisation

7.4.1.1 Overview of the Virtual Test Platform

In order to perform optimisation, a dynamic virtual test platform, which is a real-time simulation of a building and its HVAC system, is often used for analysing and evaluating different design alternatives for building HVAC systems under the simulated "real working conditions". The key features of the virtual test platform are summarised as follows.

- The test platform is developed using detailed numeric physical models to simulate the dynamic energy and thermal performance of the buildings and HVAC systems including sensors and control actuators.
- The test platform can be specified based on the building/HVAC system designs or the actual operation data of buildings/HVAC systems.
- The test platform, representing the actual systems, can be used to evaluate the building energy and operation/control performance by means of different system configurations, different component sizing and different operation/control strategies, resulting in an optimised system design.
- The test platform can be used to test and evaluate the operational reliability and energy performance of different control strategies prior to their implementation.

7.4.1.2 The Programming Software Used

The entire virtual test platform was developed using the TRNSYS (TRaNsient SYstem Simulation) program, which is a complete and extensible software for the transient simulation of systems. TRNSYS is widely used by engineers and particularly researchers to validate new energy concepts, including control strategies, occupant behaviour, alternative energy systems (wind, solar, photovoltaic, hydrogen systems), etc.

7.4.1.3 The Development of the Platform

The virtual test platform comprises of the buildings, the HVAC systems and the monitoring/control devices and control strategies.

The building is simulated using the embedded multi-zone building model (a building model included in TRNSYS), which can calculate the dynamic heat transfer through the windows and the walls under specific weather conditions. The heat transfer process covers detailed calculation of thermal conduction, convection and radiation. The material and dimension of the buildings can be specified using the information of the real building. The cooling load can be calculated after properly setting the parameters of the walls, windows, occupancy, working schedules and weather conditions.

The HVAC systems are simulated, including the components such as chillers, cooling towers, pumps, air-handling units, primary air units, the VAV (Variable Air Volume) system and hydraulic networks. Detailed or simplified physical models are used to simulate the energy and thermal performance of each component. The sizing and capacity of each component can be specified in accordance with the actual components. The hydraulic networks are developed using the detailed pressure balance models, which can accurately distribute water to each branch.

The control systems are simulated, including the sensors, the actuators and the control logics. Controllers receive the real-time operating information from the sensors, and send the command signals to modulate the control devices or the control set-points of each component, such as valve opening, the operating speed, the operating number, the supply water temperature, etc.

7.4.2 Approaches of Energy Optimisation for New Buildings

The energy optimisation of new buildings is performed as the building progresses through the design stage, the construction stage, test and commissioning stage, and the operation stage, as illustrated by Fig. 7.2.

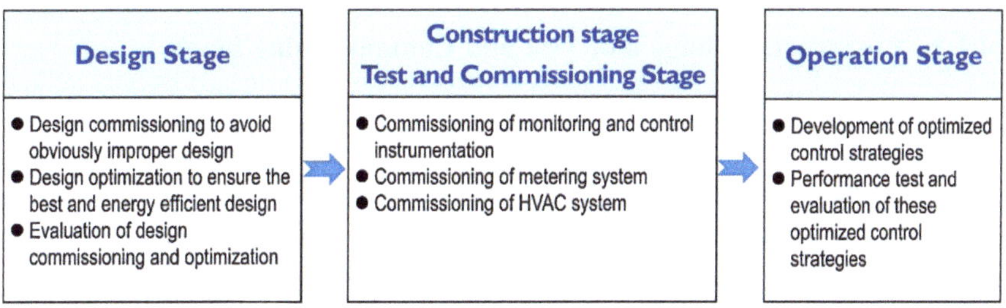

Fig. 7.2 Schematics of energy optimisation approaches in building life-cycle. *Source* Authors' work

7.4.2.1 Design Stage

Design commissioning: Commissioning in the design stage focuses on verifying the system configuration and component selection. By evaluating the designed system, any improper system configuration and component selection, in particular the type and sizing, will be identified. The major commissioning work includes:

- Verification of design configurations
- Verification of component selection and sizing

Design optimisation: The genuine objective of design optimisation here is to achieve a cost-effective option which provides the system with the possibility to the operate at high efficiency under all possible conditions, particularly at partial load, and to be monitored and controlled effectively, including the ability to absorb the uncertainties from design. The major optimisation work includes:

- Optimisation of design configurations
  E.g. chilled water distribution system optimisation (primary constant only/primary constant–secondary variable/primary variable only), connection of cooling tower (parallel/one tower follows one chiller).
- Optimising component selection and sizing
  E.g. the number and capacity of chillers, the pump head optimisation, the number and capacity of cooling towers, the air flow and cooling capacity of air handling units (AHUs).
- Optimal selection of advanced technologies
  E.g. renewable energy implementation, waste heat implementation, heat recovery; heat/cooling storage, liquid desiccant technology.

Evaluation of design commissioning and optimisation: The results of design commissioning and optimisation can be evaluated using the developed simulated platform based on the design information. The improper design alternatives found can be introduced on the test platform to identify how much energy will be wasted when compared with a better design. Similarly, compared to the original design, the energy benefit of an optimised alternative in design also can be assessed and provided.

7.4.2.2 Construction Stage and Test and Commissioning Stage

Commissioning of monitoring and control instrumentation: The verification of the installed monitoring and instrumentation is essential in ensuring the metering system functions as well as expected. Otherwise, the operation and control cannot achieve a satisfactory performance. According to the verification results, associated corrections will be made to allow the system to achieve the intended performance. Not only can massive energy/cost savings be achieved through the verification, but also the robustness of system control can be greatly enhanced.

Commissioning of metering systems: The metering system is the foundation for the system commissioning, operation and control. In the commissioning process, the major works include:

- Verifying the metering systems in terms of the type and accuracy/cost as well as installed locations.
- Proper calibration of the metering systems.
- Improving the metering systems by deleting those unnecessary sensors/meters or adding extra sensors/meters essential for energy efficient control.

Commissioning of HVAC systems: The components, sub-systems and the entire HVAC system are tested and verified to ensure their performance meets the design intent. The major works include:

- Verifying the components and system installations.
- Testing the operation performance of components, subsystems and the entire system.

7.4.2.3 Operation Stage

During operation stage, optimal control strategies are developed and implemented to ensure the entire system operates at the best possible energy efficiency, particularly under part load conditions. The major works include:

- Development of optimal control strategies for the chillers, cooling towers, chilled water systems and air-side systems.
- Performance testing and evaluation of these optimal control strategies in the virtual test platform prior to their real application.
  Optimal control strategies developed by the authors' team and that which were implemented in real buildings include:
- *Robust Chiller Sequence Control* (Sun et al. 2009)
- *Model-Based Optimal Chiller Start Control Strategy* (Sun et al. 2010)
- *Global Optimal Control of Chilled Water Systems* (Ma/Wang 2009b)
- *Optimal Sequence Control of Intermediate Heat Exchangers* (Wang et al. 2013)
- *Optimal Speed Control of Water Pumps Distributing Water to Terminals Units* (Ma/Wang 2009a)
- *Optimal Speed Control of Water Pumps Distributing Water to Heat Exchangers* (Wang et al. 2010)
- *Optimisation of Supply Water Temperature Set-point at Secondary Side of Heat Exchangers* (Wang et al. 2013)
- *Demand Side Flow-Limiting Control Strategy* (Gao et al. 2011)
- *Global Optimisation of Cooling Tower Systems* (Ma et al. 2009)
- *Ventilation Control Strategy for Multi-zone Air-Conditioning Systems* (Sun et al. 2011a, b)

- *Static Pressure Set-Point Reset* (Wang 2010)
- *Supply Air Temperature Set-Point Reset* (Wang 2010)

7.4.3 Approaches of Energy Optimisation for Existing Buildings

For existing buildings, the energy optimisation aims to find out the possible energy saving opportunities by performance assessment and fault diagnosis. The related energy saving measures are then offered and implemented to realise the energy savings in practical operation. The energy optimisation work can be carried out in three stages.

7.4.3.1 Stage 1: Preliminary Building Energy Assessment and Identification of Energy-Saving Areas

The objective of the work at this stage is to provide a preliminary energy performance assessment, make a basic picture of the possible energy saving potential, and provide preliminary energy-saving options. The main work and deliverables include:

- Collecting basic information of the cooling system, such as system configuration/sizing, operation and control strategy, history operation data (electricity consumption, water flow, water/air temperature, pressure, etc.). Some operation data are essential for performance assessment, such as the actual coefficient of performance of chillers, the operating head of pumps, the temperature difference of the overall chilled water system, the speed of pumps and AHU fans, etc. If these essential data can be provided, some onsite measurements will be conducted.
- Conducting preliminary energy performance analysis and assessment based on the data collected.
- Providing a report to estimate preliminary energy-saving potentials as well as the associated energy saving options, based on the analysis and diagnosis results.

7.4.3.2 Stage 2: Detailed Energy Performance Diagnosis and Evaluation of Alternative Energy-Saving Measures

The objective of the work at this stage is to provide alternative energy-saving measures to be implemented based on the detailed diagnosis, cost-effectiveness assessment and practical constraints. The main work and deliverables include:

- Collecting detailed system design information and large amounts of operation data.
- Conducting detailed site measurements to get more essential data (if needed).
- Conducting detailed energy diagnosis and assessment to find out the specific energy-saving opportunities.
- Providing alternative energy-saving measures in terms of fine-tuning, retrofitting and even upgrading the equipment/subsystem/control system, and clearly assessing the energy-saving benefit of each measure.
- Providing alternative control strategies optimised for the specific project, and clearly assessing the energy-saving benefit of each control strategy.

7.4.3.3 Stage 3: Implementation of the Energy Optimisation Measures Proposed

The objective of the work at this stage is to implement the energy optimisation measures proposed into the practical building systems. The main work and deliverables include:

- Providing the site implementation procedures for tuning/retrofitting/upgrading of the equipment/subsystem/control system.
- Developing the software package for the optimal control strategies.
- Implementing the optimal control strategies in building automation systems.

7.4.4 Energy Performance Assessment Method for Information-Poor Buildings

There exist many information-poor buildings (Dexter 2012). The major characteristics of such buildings are summarised as follows.

- Insufficient energy consumption data:
  - Only the total building electricity is available based on the monthly electricity bill.
  - There are few or no sub-meters for different sub-systems.
- Poor-quality measurements:
  - There are very limited sensors used for performance monitoring.
  - The measuring errors are very large due to a lack of calibration and maintenance.

- Unknown, abnormal operation modes:
  - The occupancy behaviours are not known.
  - Failures of automatic control result in large deviations from normal operations.

The question of how to obtain required performance data based on limited available information is the biggest challenge for assessing the energy performance in information-poor buildings.

A simplified energy performance assessment method based on macroscopic energy balance principles is developed for information-poor buildings (Yan et al. 2012). Though it requires very limited building energy data, it can effectively assess the energy performance at building and system levels, and disaggregate the whole-building consumption into consumption by three groups of end-users—the HVAC consumers; internal-consumers; and other-consumers.

7.4.4.1 Basic Concept

Figure 7.3 shows two typical energy flows in an air-conditioned building, which can be used to determine the energy performance of different systems. The first energy flow is the electrical energy flow that depicts electricity consumption of all end-users in the building. All these end-users are classified into three categories in

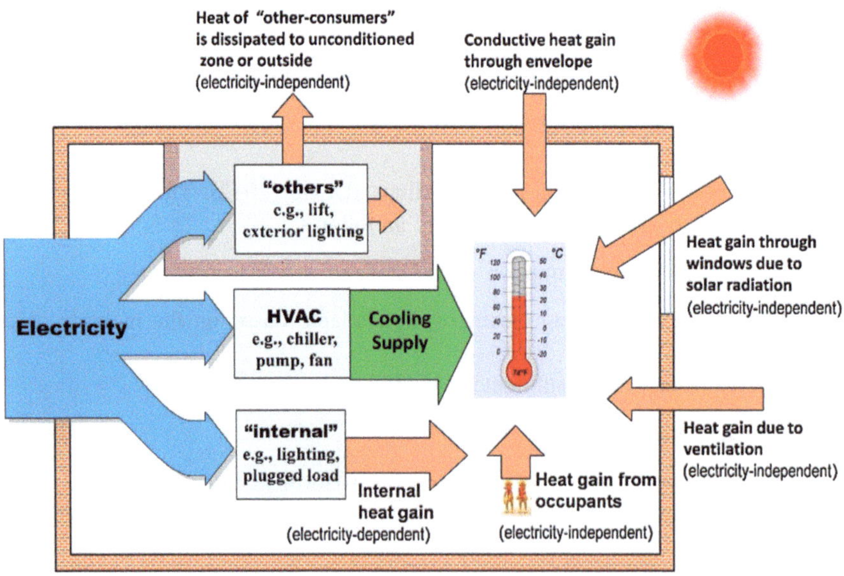

Fig. 7.3 Electrical and thermal energy flows in an air-conditioned building. *Source* Authors' work

this study: the A/C system; "internal-consumers" (e.g. lighting system, indoor equipment and appliances); and "other-consumers" (e.g. lift, outdoor lighting). The second energy flow is the thermal energy flow that describes all heat transfer processes in the building. These heat transfer processes can be classified as electricity-dependent (e.g. the associated heat dissipation of electricity consumers) and electricity-independent (i.e. various external heat gains and the heat gain of occupants) heat transfer processes. The electrical energy flows of these three consumers have very different impacts on the thermal energy flows in the air-conditioned space. The A/C system produces cooling energy to the building by consuming electricity. The "internal-consumers" almost release all dissipated heat into the air-conditioned area as internal heat gains. The energy consumed by "other-consumers" almost has no impact on the heat balance in the air-conditioned area since the dissipated heat is released into the unconditioned area or is exhausted to the outside directly.

Based on the two energy flows, two energy balances can be established for an air-conditioned building.

Two overall energy balances exist in an air-conditioned building. The first is the electricity energy consumption balance at the whole building level as expressed in Eq. (7.1), which is used to break down the total energy consumption into the individual consumption of three systems

$$E_{Building} = E_{A/C} + E_{Internal} + E_{Others} \tag{7.1}$$

where, $E_{Building}$ is the total energy consumption of the whole building, which is given by the monthly energy bills. $E_{A/C}$, $E_{Internal}$ and E_{Others} are the monthly energy consumption of the A/C system, "internal-consumers" and "other-consumers", respectively.

The second is the cooling energy balance between the demand side and supply side of the A/C system as expressed in Eq. (7.2), which is used to determine the energy performance indicators of the A/C system and main components

$$CL_{Supply} = CL_{Demand} \tag{7.2}$$

where, CL_{Demand} and CL_{Supply} are the monthly cooling load calculated at the demand side and supply side, respectively.

An optimisation algorithm using the trial-and-error method is developed to search the best parameters for the trade-off of the balance equations. The principle of this algorithm is that the optimised parameters drive the best possible cooling energy balances between demand side and supply side and still maintain the electricity consumption balance simultaneously. The optimised parameters are actually the energy consumption of three systems that lead to the minimal imbalance between the supply side and the demand side of the cooling energy balance.

7.4.5 Examples of Technologies/Tools Developed

Based on the above methods of building energy optimisation, a series of technologies/tools are developed. Some examples are shown as below.

7.4.5.1 Building Energy Commissioning/Diagnosis Tools

- Building Level Energy Performance Quick Evaluation and Diagnostic Tool
- Detailed Evaluation and Diagnostic Tool for A/C and BA systems.

7.4.5.2 Virtual Test Platforms

- Building System Online Performance Simulation Test Platform
- BA Control and Diagnosis Strategy Online Test Platform.

7.4.5.3 Optimisation Tools

- Building HVAC System Optimisation Tool
- Package of Online Optimal and Energy Efficient Control and Fault Diagnosis Strategies.

7.5 On-site Implementation of the Control Strategies Developed

For on-site application issues, the developed optimal control strategies should be developed into software packages that are compatible with the building automation system (BAS) used in the building. An easy solution is to program the developed control logic under the framework of the specific BAS used in this building.

Alternatively, one can program and implement the developed control logics under the framework of a third-party communication platform. The third-party platform refers to the communication platform that has good compatibility and is able to exchange information with different BASs.

The following parts of this section present an example of an open BMS (Building Management System) communication platform and how to develop and implement user-defined control modules for real applications.

7.5.1 An Example of Third-Party Management and Communication Platform

The management and communication platform, namely IBmanager (Wang et al. 2005), was developed by the authors' team in the Hong Kong Polytechnic University, as shown in Fig. 7.4. IBmanager is an open BMS management and communication platform based on middleware and web services technologies to support the integration and management of building automation systems from different vendors as well as remote monitoring and management services. The IBmanager uses the BACnet protocol which is proposed by ASHRAE standard 135-2008.

7.5.2 Implementation of Control Strategies via IBmanager

The developed control strategies were programmed using the application program of Matlab and compiled as dynamic link library (DLL) modules that can be

Fig. 7.4 Interface of IBmanager. *Source* Screen capture from IBmanager homepage

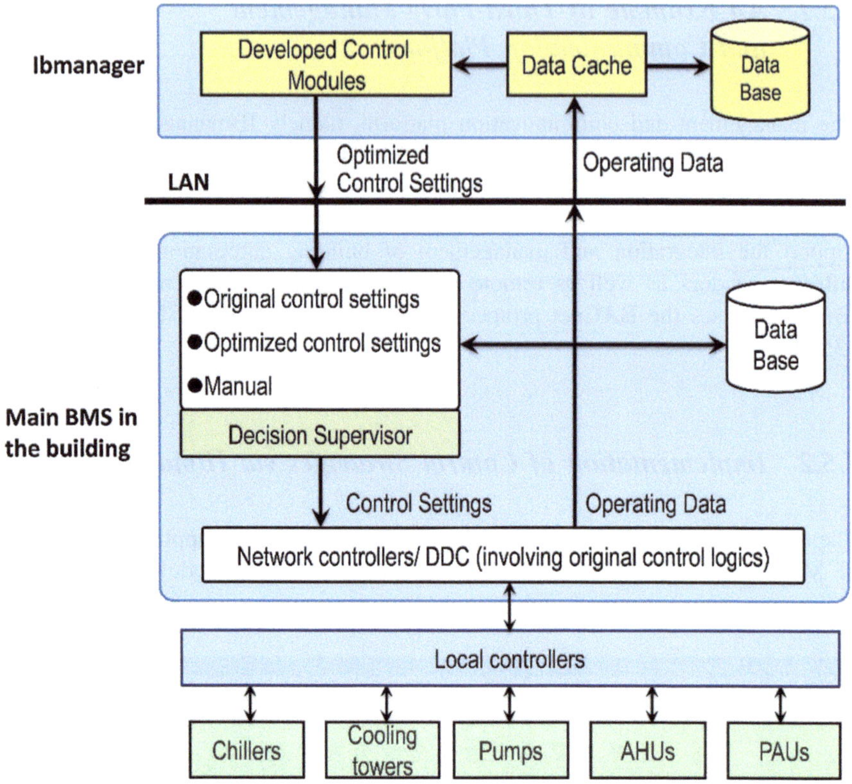

Fig. 7.5 Implementation structures of developed control strategies. *Source* Authors' work

integrated with IBmanager. IBmanager, involving the developed control modules, runs on a separated PC (Personal Computer) station connected with the building BMS platform.

As shown in Fig. 7.5, the operating data of HVAC systems (e.g. chillers, cooling towers, pumps) will be collected by the network controllers or DDC (direct digital controller) from the local controllers. IBmanager can receive these data through LAN (local area network), and then send these data to the developed control modules. After calculations, optimised control settings will be produced by the developed modules and sent back to the decision supervisor of the main BMS in the building. The decision supervisor allows the operators to determine whether the optimised settings given are used or ignored (not used). The final control settings will be sent to the local logics used in the building to achieve energy efficient control of HVAC systems through network controllers/DDC and local controllers.

7.6 Practice of Energy Optimisation in a New Building: A Super High-Rise Building

7.6.1 Descriptions of the Building and Its Central Cooling System

The International Commerce Centre, a 118-storey and 490-m super high-rise commercial building, is currently the tallest building in Hong Kong. It has a gross floor area of approximately 321,000 m^2 excluding the hotel at the top of the building. It serves as a commerce centre comprising commercial offices, shopping arcades and a six-star hotel.

A central chilling system (shown in Fig. 7.6) provides cooling to the building spaces; it consists of six identical centrifugal chillers (each with a cooling capacity of 7230 kW) and 11 indoor cooling towers (each with a motor power of 150 kW). Plate heat exchangers are employed to deliver cooling from low levels to high floors to avoid extremely high static pressure. Each chiller is interlocked with a constant condenser water pump and a constant primary chilled water pump. All pumps in the secondary chilled water distribution system are equipped with variable frequency

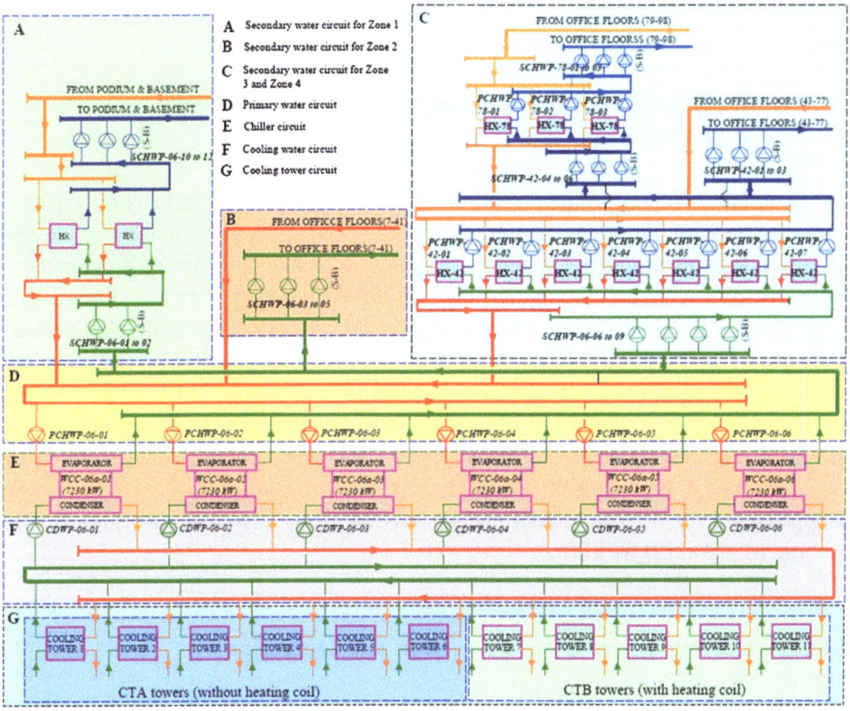

Fig. 7.6 Central cooling system layout. *Source* Ma et al. (2008b)

drivers (VFD) except that the primary chilled water pumps dedicated to the heat exchangers in Zones 3 and 4 are constant speed pumps. Each office floor in this building is served by two AHU systems involving variable air volume (VAV) Boxes.

Since the HVAC system design stage (2005) and after the first stage of occupation in 2008, the authors' team, together with the developer, consultant, facility management team and contractors, have made serious efforts to develop innovative solutions for enhancing energy efficiency and improving environmental performance. The innovative solutions implemented mainly include improved system configurations, optimised control strategies and proper maintenance. In 2012, 2013 and 2014, the ICC management services office was awarded the "Gold" Class of LOOP (Low-carbon Office Operation Program); the "Platinum" grade of LEED CI (Leadership in Energy and Environmental Design—Commercial Interior); and the ASHRAE Technology Award, respectively.

7.6.2 *Examples of Design Optimisation*

Design commissioning and optimisation mainly concern the future operation and control performance of HVAC systems. Some examples of design optimisation are illustrated as follows.

7.6.2.1 Simplification of the Secondary Water Loop Systems of 3rd/4th Zones

The original designed primary–secondary pumping paradigm in the upper part of the building (3rd/4th zones) employed primary constant speed pumps at the secondary side of each heat exchanger. Actually, the major function of these primary pumps is to provide the circulation force of chilled water to overcome the pressure drop of heat exchangers. Under part load, especially under light load, these primary chilled water pumps contribute a large portion of the total energy used in the 3rd/4th zones. To reduce initial cost and save operating energy, the original design was revised (Ma et al. 2008a), thus eliminating the primary constant speed pumps, as shown in Fig. 7.7. The on-site tests showed that the power consumption of the variable speed pumps for Zone 3 was slightly increased while the total pump energy dropped by 55.8 %. The annual energy savings using this revised design configuration are over 1.0 M kWh.

7.6.2.2 Cooling Tower System Selection and Operation

The original design of the cooling tower system employed two-speed (two-stage) fans (and later replaced by variable speed fans with the minimum allowed speed of

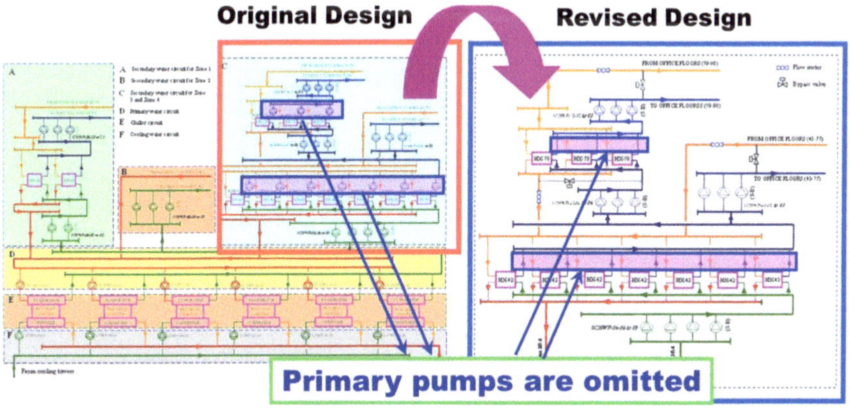

Fig. 7.7 Central cooling system layout. *Source* Ma et al. (2008a)

37 Hz). In order to maximise the energy savings, cooling tower fans were revised to variable speed using VFD. In operation, a reduction in minimum operating frequency of fans from 37 to 20 Hz was implemented, which was confirmed to be feasible and reliable via simulation and then tests. Based on the commissioning test results, such use of VFD and lower operating frequency provide an annual energy saving of up to 2.36 million kWh.

7.6.3 *Examples of Optimised Control Strategies Implemented*

7.6.3.1 Robust Chiller Sequencing Control

One conventional chiller sequencing control method is to determine the number of chillers required according to the measured building cooling load. However, the direct measurement of the building cooling load might be inaccurate or unreliable because of uncertainties associated with the measurement instrument. The ICC uses data fusion to improve the reliability of building cooling load measurement, while providing a degree of confidence in the fused load measurement for robust control and fault detection, as shown in Fig. 7.8 (Sun et al. 2009).

The advantages of direct measurement and indirect measurement are merged in the final fused cooling load measurement. A chiller maximum cooling capacity online calculation and data fusion are used to improve the reliability of chiller sequencing control. Test results show about half of the chiller switch operations were avoided and that more than 1 % (680,000 kWh per year) of the chiller plant energy consumption can be saved.

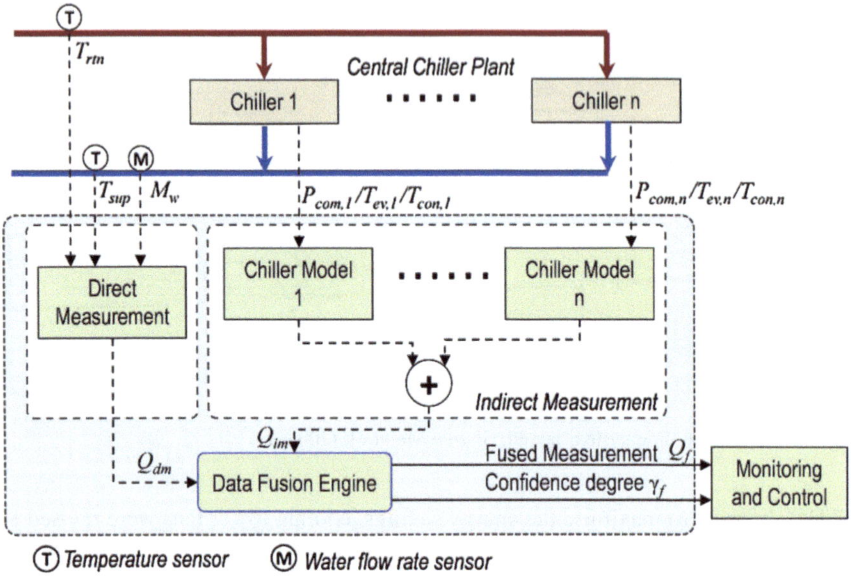

Fig. 7.8 Data fusion used for cooling load measurement. *Source* Sun et al. (2009)

7.6.3.2 Optimal Cooling Tower Control

To achieve energy savings under low partial load conditions, a hybrid quick search (HQS) method (Ma et al. 2009) was developed and implemented to minimise the instantaneous total power consumption of the chillers and the cooling tower fans. The required number of operating cooling towers and the water temperature entering the condensers are continuously reset based on weather conditions and system working conditions. In situ test results confirm a 1–4 % energy saving of central plant (cooling towers and chillers) consumption is achieved using this strategy, as compared to using a fixed approach.

7.6.3.3 Optimal Control of the Secondary Water Pumps

A cascade controller, as shown in Fig. 7.9, is used to control the operating speeds of pumps distributing water to heat exchangers (fully open the control valves) instead of using the modulating valves while keeping a fixed differential pressure (Wang and Ma 2010). The fully opened valve in the cascade control minimises the water loop resistance and therefore saves the energy of pumps. The test results proved good control reliability of the strategy as well as annual energy savings of up to 250,000 kWh.

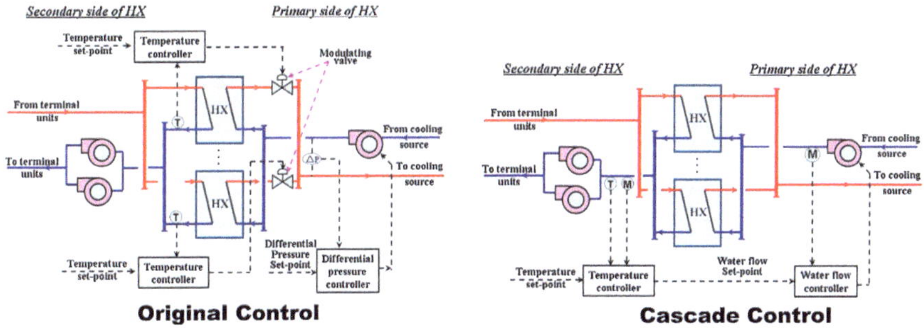

Fig. 7.9 Comparison of the original control and the cascade control. *Source* Wang/Ma (2010)

7.6.3.4 DCV and Model-Based Outdoor Air Ventilation Control

To solve over-ventilation or under-ventilation found in conventional ventilation controls, a robust DCV control strategy based on ASHRAE standard 62.1 was developed and implemented (Sun et al. 2011a, b). Additionally, a model-based free cooling outdoor air control strategy was combined with the DCV strategy. The test results from commissioning and operation confirm that energy consumption of the primary air-handling unit (AHU) can be reduced up to 50 % under part load compared with the original two-stage control, and energy consumption for cooling outdoor air can be reduced by up to 65 %. The CO_2 concentration was maintained below 800 ppm in each zone for good IAQ.

7.6.3.5 Indoor Air Quality and Thermal Comfort

For the typical office floor in this building, an individual air-handling unit usually provides cooled air to Variable Air Volume (VAV) boxes located in multiple separated zones. A robust demand controlled ventilation (DCV) control for multi-zone office floors has been developed and implemented (Sun et al. 2011a, b), which is based on ASHRAE standard 62.1. The innovative control method predicts the real-time occupancy for determination of the required total fresh air. Under the conditions of low ventilation demand, this method can significantly reduce the energy consumed to cool the additional fresh air. In situ test results show that the measured CO_2 concentrations of zones are always under 800 ppm (shown in Fig. 7.10), a good indoor air quality. The measured indoor air temperatures of typical floors are well controlled between the comfort range (i.e. 23–26 °C) during office hours (shown in Fig. 7.11).

Other optimal control strategies used for energy efficiency of the HVAC system include:

- Flow-limiting control for eliminating deficit flow, avoiding low delta-T syndrome and saving pump energy (Gao/Wang 2011).
- Chilled water supply temperature optimisation (Ma/Wang 2009a, b).

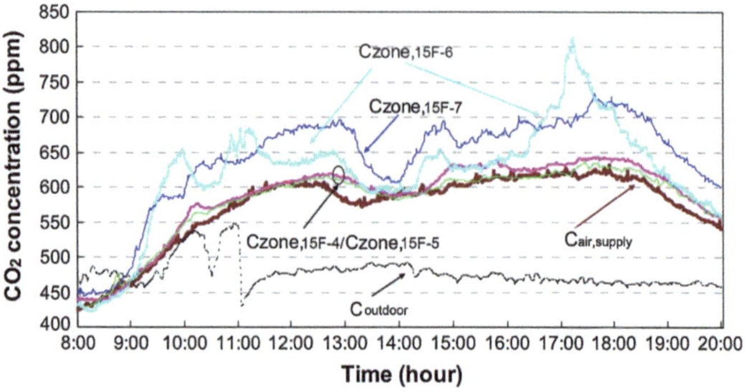

Fig. 7.10 Measured indoor air CO_2 concentration of individual zones at 15/F. *Source* Authors' calculation

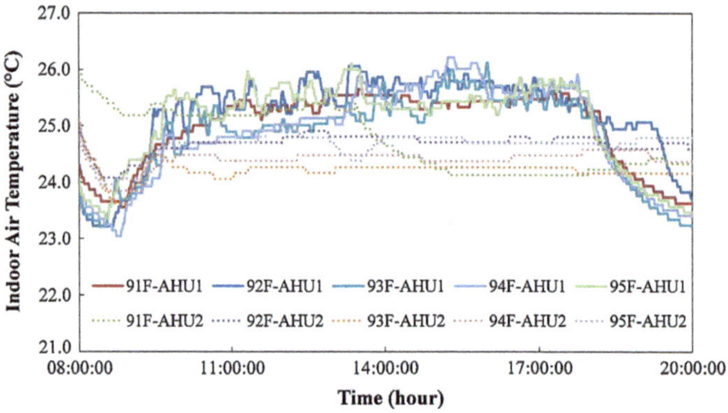

Fig. 7.11 Room temperature of five typical floors (91st to 95th). *Source* Authors' calculation

- AHU supply air temperature optimisation.
- Differential pressure set-point optimisation in the secondary pumps (Ma/Wang 2009a, b).

7.6.4 Energy Benefits

The projected and actual energy performance is compared with the baseline performance based on ASHRAE 90.1-2007, as summarised in Table 7.1. The projected building energy use intensity (EUI) is 182.5 kWh/m^2/year, representing a 30.2 % energy saving compared with the baseline performance, i.e. EUI of 261.4 kWh/m^2/year. The actual EUI in 2012 is 164.5 kWh/m^2/year, representing a 37.1 % energy saving compared with the baseline performance.

Table 7.1 Comparison of actual building energy consumption year-round

	Baseline	Projected	Actual (2012)
Total energy consumption (kWh)	83,898,000	58,582,000	52,805,000
Energy use intensity (EUI) (kWh/m^2)	261.4	182.5	164.5
Energy savings	–	30.2 %	37.1 %

Source Authors' calculation

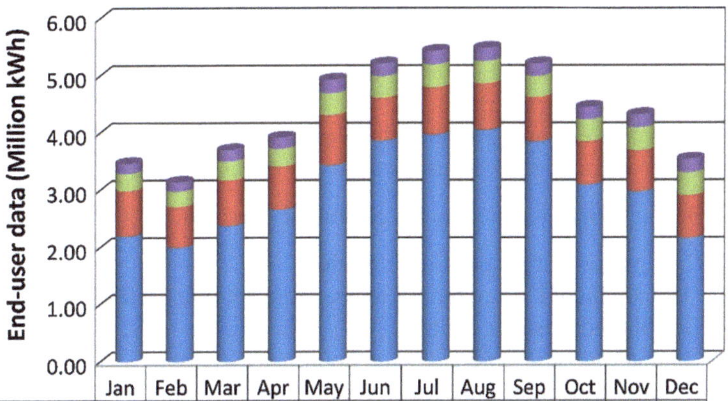

	Jan	Feb	Mar	Apr	May	Jun	Jul	Aug	Sep	Oct	Nov	Dec
Other	0.18	0.16	0.20	0.20	0.24	0.22	0.24	0.23	0.22	0.22	0.23	0.24
Lift	0.30	0.27	0.33	0.31	0.38	0.38	0.40	0.39	0.37	0.38	0.40	0.40
Lighting&Equipment	0.79	0.71	0.79	0.75	0.88	0.75	0.82	0.81	0.77	0.76	0.72	0.74
HVAC	2.20	2.00	2.38	2.66	3.43	3.86	3.97	4.05	3.85	3.09	2.97	2.16

Fig. 7.12 Monthly energy consumption (million kWh) of different systems in 2012. *Source* Authors' calculation

Figure 7.12 shows the actual monthly energy consumption of different service systems of the building in 2012. Compared with the EUI of typical office buildings in Hong Kong (i.e. 270 kWh/m^2), the energy saving of ICC in 2012 was about 39.1 %.

Compared with the baseline case based on ASHRAE 90.1, the innovative technologies used result in an annual cost saving about HKD 30 M (million). Compared with the original HVAC system configuration/design and conventional control strategies originally developed for the HVAC system, the annual cost saving of about HKD 7 M is achieved through the application of the life-cycle commissioning and optimisation. For implementation of the developed optimal control strategies, extra hardware investment is about HKD 2 M while there is no extra cost for implementing the HVAC system design optimisation. Additionally, HKD 3 M of manpower cost was spent on the strategy development and implementation. With HKD 7 M cost saving in the HVAC system, the actual payback period is less than one year, i.e. 8.6 months.

7.7 Practice of Energy Optimisation in New Buildings: A Hotel Building

Building system commissioning and optimisation have been implemented in another new real building: a hotel building in Hong Kong. It has a total floor area of 4477.144 m^2. The central chilling system is a typical variable-primary chilled water system, including two identical screw chillers (190 tons each).

The system design diagnosis and optimisation include:

- Bypass of heat recovery in fresh air supply system
- Reset of the cooling water system design pressure drop
- Optimal chiller selection
- Optimal condensing water pump selection
- Optimal cooling tower installation

Optimal control strategies for central air-conditioning systems include:

- Cooling tower fan speed optimisation
- Optimal chilled water supply temperature set-point reset strategy
- DCV-based fresh air control
- Optimal pressure differential set point reset strategy
- Optimal condenser inlet water temperature set point reset strategy
- Optimal cooling tower sequence control
- Optimal cooling water pump sequence control
- Optimal PAU fan speed control
- Optimal chiller sequence control

Through commissioning (improving the system configuration and selection) and control optimisation, the energy savings of the whole HVAC system are about 20 % in a year, compared with when the HVAC system operated correctly as the original design intent.

7.8 Practice of Energy Performance Assessment in Existing Information-Poor Buildings

7.8.1 Software Tool: Energy Performance Assessment and Diagnosis for Information-Poor Buildings

A software tool was developed for practically implementing the energy performance assessment method for information-poor buildings, as shown in Figs. 7.13 and 7.14. The software consists of three function modules: the input module, assessment module and output module.

In the input module, the required data mainly include monthly electricity bills, general building design data, weather conditions and the (very limited) design and

Fig. 7.13 Main interface of the software. *Source* Screen capture from software homepage

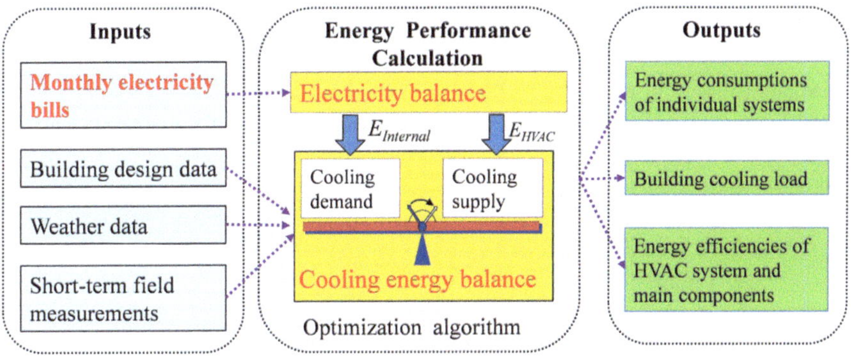

Fig. 7.14 Schematic of the developed assessment software tool. *Source* Yan et al. (2012)

operation data of the HVAC system. Apart from the weather data that may be provided by the observatory, other data can be obtained from the already available documents or supplemented by short-term in situ measurements. The developed energy performance assessment method above, which is the core part of the software tool, is implemented in the assessment module. The outputs of the tool include the disaggregated energy consumption of the three systems, and the energy performance indicators of the HVAC system on both the demand side (e.g. building cooling load) and the supply side (e.g. energy efficiency).

7.8.2 Application Case

The developed software tool was applied in an office building in Hong Kong, a super high-rise building, which consists of a basement of four floors, a block building of six floors (a commercial centre), and a tower building of 98 floors (commercial offices).

In this building, the required input data were mainly collected from available documents, such as the payments of monthly electricity bills, the building design documents and manufactory manuals. The important operating data of the HVAC system were obtained from the operation records and supplemented by the in situ measurements in a typical cooling day. All required input data are summarised in Table 7.2.

The energy performance data of the building at multiple levels are shown in Fig. 7.15 and Table 7.3. The total energy consumption of the whole building is

Table 7.2 Required inputs for energy performance assessment

Data group	Data description
Electricity bill data	Monthly energy consumption data of the whole building
Basic design data	Area, U-factor and SC value of building envelopes
	Design value of the indoor temperature, RH, fresh air rate and occupant
Basic operating data	The timetable of building operating, the number of running chillers
Field measurements in a typical cooling day	Building cooling load data
	Power consumption of the HVAC system and main component

Source The authors

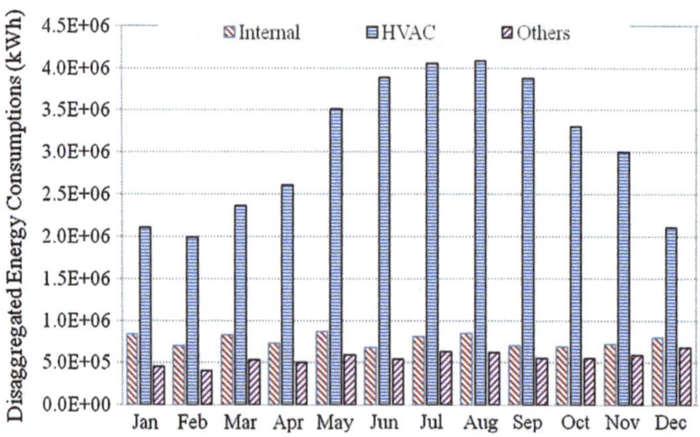

Fig. 7.15 Disaggregated energy consumption of three systems. *Source* Authors' calculation

Table 7.3 Estimated monthly energy performance indicators of the HVAC system

Month	Cooling load (kWh)	Overall energy efficiency (SCOP)	α (%)
Jan	2,662,865	1.26	37.5
Feb	2,577,273	1.29	37.8
Mar	3,288,967	1.42	36.6
Apr	4,829,734	1.79	33.8
May	7,680,527	2.11	32.0
Jun	8,285,003	2.19	31.6
Jul	9,354,749	2.22	31.4
Aug	9,848,695	2.38	30.7
Sep	8,924,187	2.27	31.2
Oct	7,031,887	2.10	32.0
Nov	5,638,667	1.91	33.1
Dec	3,284,464	1.50	35.9

Source Authors' calculation

disaggregated into the individual consumptions of three systems (i.e., the HVAC, internal-consumers and other-consumers), which are the most important data for assessing the energy performance at the system level. The building cooling load and the overall energy efficiency (SCOP) were mainly used to assess the energy performance of the HVAC system on the cooling demand side and the cooling supply side respectively. $\boldsymbol{\alpha}$ can be used to assess the energy performance of the cooling delivering system. $\boldsymbol{\alpha}$ represents the heat gain ratio of the air-conditioning system, which indicates that a part of the energy use of the air-conditioning system can be converted into an additional heat gain during the cooling energy transport processes.

References

ASHRAE (2008) *2008 ASHRAE Handbook—HVAC Systems and Equipment*. Atlanta: American Society of Heating, Refrigerating and Air-Conditioning Engineers, Inc.

Bordass, Bill, Robert Cohen, Mark Standeven and Adrian Leaman (2001) "Assessing Building Performance in Use 3: Energy Performance of the Probe Buildings". *Building Research & Information* 29, 2: 114–28.

Claridge, D., M. Liu and J. Haberl (1996) "Can You Achieve 150 % of Predicted Retrofit Savings? Is It Time for Recommissioning?". Proceedings of 1996 ACEEE Summer Study on Energy Efficiency in Buildings, August, pp. 4.59–4.67.

Claridge, D., C. Culp, M. Liu et al. (2000) "Campus-Wide Continuous Commissioning of University Buildings". Proceedings of 2000 ACEEE Summer Study on Energy Efficiency in Buildings. Pacific Grove, CA, August 20–25, pp. 3.101–3.112.

Dexter, A.L. (2012) *Monitoring and Control of Information-Poor Systems: An Approach based on Fuzzy Relational Models*. Chichester: John Wiley & Sons Ltd.

DOE, US Department of Energy. (2012) *2011 Buildings Energy Data Book*. Washington, DC: US Department of Energy.

Hull, D.G. (2003) *Optimal Control Theory for Applications*. New York: Springer-Verlag.

Gao D.C., S.W. Wang and Y.J. Sun (2011) "A Fault-Tolerant and Energy Efficient Control Strategy for Primary–Secondary Chilled Water Systems in Buildings". *Energy and Buildings* 43, 12: 3646–56.

Lam, J.C., Y.C. Chan, C.L. Tsang and H.W. Li (2004) "Electricity Use Characteristics of Purpose-Built Office Buildings in Subtropical Climates". *Energy Conversion and Management* 45: 829–44.

Liu, M., A. Athar and D. Claridge et al. (1994) "Reducing Building Energy Costs using Optimized Operation Strategies for Constant Volume Air Handling Systems". Proceedings of the Ninth Symposium on Improving Building Systems in Hot and Humid Climates. Arlington, Texas, May 19–20, pp. 192–204.

Ma, Z.J. and S.W. Wang (2009a) "Energy Efficient Control of Variable Speed Pumps in Complex Building Central Air-Conditioning Systems". *Energy and Buildings* 41, 2: 197–205.

Ma, Z.J. and S.W. Wang (2009b). "An Optimal Control Strategy for Complex Building Central Chilled Water Systems for Practical and Real-Time Applications". *Building and Environment* 44, 6: 1188–98.

Ma, Z.J., S.W. Wang and W.K. Pau (2008a) "Alternative Designs of the Complex Secondary Chilled Water System for a Super High-Rise Building". *ASHRAE Journal* 50, 5: 42–52.

Ma, Z.J., S.W. Wang and W.K. Pau (2008b) "Secondary Loop Chilled Water in Super High-Rise". *ASHRAE Journal* 50, 5: 42–52.

Ma, Z.J., S.W. Wang and F. Xiao (2009) "Online Performance Evaluation of Alternative Control Strategies for Building Cooling Water Systems Prior to In-Situ Implementation". *Applied Energy* 86, 5: 712–21.

Nassif, N., S. Moujaes and M. Zaheeruddin (2008) "Self-Tuning Dynamic Models of HVAC System Components". *Energy and Buildings* 40: 1709–20.

Omer, A.M. (2008) "Energy, Environment and Sustainable Development". *Renewable and Sustainable Energy Reviews* 12, 9: 2265–300.

Perez-Lombard, L., J. Ortiz and C. Pout (2008) "A Review on Buildings Energy Consumption Information". *Energy and Buildings* 40, 3: 394–8.

Sun, Y.J., S.W. Wang and G.S. Huang (2009) "Chiller Sequencing Control with Enhanced Robustness for Energy Efficient Operation". *Energy and Buildings* 41, 11: 1246–55.

Sun, Y.J., S.W. Wang and G.S. Huang (2010) "Model-Based Optimal Start Control Strategy for Multi-Chiller Plants in Commercial Buildings". *Building Services Engineering Research and Technology*. 31(2): 113–29.

Sun, Z.W., S.W. Wang and N. Zhu (2011a) "Model-Based Optimal Control of Outdoor Air Flow Rate of an Air-Conditioning System with Primary Air Handling Unit". *Indoor and Built Environment* 20, 6: 626–37.

Sun, Z.W., S.W. Wang and Z.J. Ma (2011b) "In-Situ Implementation and Validation of a CO_2-Based Adaptive Demand-Controlled Ventilation Strategy in a Multi-Zone Office Building". *Building and Environment* 46, 1: 124–33.

Wang, S.W. (2002) "Automatic Diagnosis and Commissioning of Central Chilling Systems". Proceedings of the International Conference on Advances in Building Technology. Hong Kong, pp. 173–80.

Wang S.W. (2010) *Intelligent Buildings and Building Automation*. London and New York: Spon Press (Taylor & Francis).

Wang, S.W. and Z.J. Ma (2010) "Control Strategies for Variable Speed Pumps in Super High-Rise Building". *ASHRAE Journal* 52, 7: 36–43.

Wang, S.W., Z.Y. Xu, J.N. Cao and J.P. Zhang (2005) "A Middleware for Web Service-Enabled Integration and Interoperation of Intelligent Building Systems". *Automation in Construction* 16, 1: 112–21.

Wang, S.W., Q. Zhou and F. Xiao (2010) "A System-Level Fault Detection and Diagnosis Strategy for HVAC Systems Involving Sensor Faults". *Energy and Buildings* 42, 4: 477–90.

Wang, S.W., D.C. Gao, Y.J. Sun and F. Xiao (2013) "An Online Adaptive Optimal Control Strategy for Complex Building Chilled Water Systems involving Intermediate Heat Exchangers". *Applied Thermal Engineering* 50, 1: 614–28.

Xiao, F., S.W. Wang and J.P. Zhang (2006) "A Diagnostic Tool for Online Sensor Healthy Monitoring in Air-Conditioning System". *Automation in Construction* 15, 4: 489–503.

Yan C.C., S.W. Wang and F. Xiao (2012) "A Simplified Energy Performance Assessment Method for Existing Buildings based on Energy Bill Disaggregation". *Energy and Buildings* 55: 563–74.

Yik, F.W.H. (2008) "Chilled Water Circuit Designs for In Situ Chiller Performance Measurement". *Building Services Engineering Research and Technology* 29, 2: 107–18.

National University of Singapore

From modest beginnings as a medical college founded in 1905 with just 23 students, the National University of Singapore is now the country's flagship university, and consistently ranks among the top two universities in Asia and top 25 in the world. It presently spans three campuses, enrols 38,000 students from more than 100 countries in many critical fields of study, and forges partnerships around the world. A global thought leader with a distinctive Asian perspective, the University continues to serve Singapore and society, nurturing talents and improving lives with innovative education and impactful research.
For more information, visit http://www.nus.edu.sg.

B. Su and E. Thomson (eds.), *China's Energy Efficiency and Conservation*,
SpringerBriefs in Environment, Security, Development and Peace 30,
DOI 10.1007/978-981-10-0737-8

Energy Studies Institute

Since its establishment in November 2007, the Energy Studies Institute (ESI) is well on its way towards becoming a presence in the international energy policy research arena. In 2015, ESI ranked 8th in the "Top Energy and Research Policy Think Tanks" category of the *Global Go To Think Tank Index Report* by the University of Pennsylvania's Think Tank and Civil Societies Program. This Institute provides timely and quality analyses in an evolving energy research landscape influenced by regional and global events. ESI's research has been published in key internationally refereed journals and other publications which include the *ESI Bulletin* that is widely read. Its researchers often write newspaper commentaries and appear on television interviews.

As of 2015, ESI has six research tracks: Energy and Climate Change; Energy and Economic Competitiveness; Energy Efficiency and Conservation; Energy for Power Generation; Energy for Transport; and Energy Geopolitics.

For more information, visit: http://www.esi.nus.edu.sg.

B. Su and E. Thomson (eds.), *China's Energy Efficiency and Conservation*,
SpringerBriefs in Environment, Security, Development and Peace 30,
DOI 10.1007/978-981-10-0737-8

About the Editors

Dr. Bin SU is Senior Fellow at the Energy Studies Institute (ESI), National University of Singapore. He joined ESI in 2011 and leads the energy modelling program at ESI. His research interests include energy and environmental systems modelling and analysis, with a particular emphasis on energy efficiency, climate change, embodied energy/emissions, carbon footprints, carbon pricing and rebound effects. He is presently Associate Editor of *Energy Economics,* Editorial Board Member of *Reports on Economics and Finance,* and has published widely in internationally refereed journals.

Address: Dr. Bin Su, Energy Studies Institute, National University of Singapore. 29 Heng Mui Keng Terrace, Block A, #10-01, Singapore 119620.
Email: subin@nus.edu.sg; subin.nus@gmail.com
Website: http://www.esi.nus.edu.sg/about-us/our-researchers/su-bin.

Dr. Elspeth THOMSON is a consultant based in Victoria, BC, Canada. She was formerly Senior Fellow and Head of the Energy and Environment Division at the Energy Studies Institute (ESI), National University of Singapore. In the 1990s she taught at Simon Fraser University in Canada and Lingnan University in Hong Hong. Her research interests span Asian energy security, energy economics, and energy and the environment. Besides her book on China's coal industry, she has written numerous articles and book chapters on many aspects of the various types of energy produced, consumed and traded in Southeast, North and South Asia.

Address: Dr. Elspeth Thomson, Victoria, BC, Canada.
Email: e_bt_56@hotmail.com

B. Su and E. Thomson (eds.), *China's Energy Efficiency and Conservation*, SpringerBriefs in Environment, Security, Development and Peace 30,
DOI 10.1007/978-981-10-0737-8

About the Authors

Dr. Ying FAN is Professor and Director of the Center for Energy and Environmental Policy Research at the Institute of Policy and Management, Chinese Academy of Sciences. She is also Vice President of the Chinese Society of Optimization, Overall Planning and Economic Mathematics, and was elected as a member of the International Association for Energy Economics Council in 2013. Her research interests are energy-environment-economy system modelling, and energy and environmental policy. She has carried out over 40 research projects, and published eight books and over 90 papers in internationally refereed journals.

Dr. Hong HUO is Associate Professor in the Institute of Energy, Environment and Economy at Tsinghua University, China. She was previously a postdoctoral Research Associate at Argonne National Laboratory. Her research interests include modelling of air pollution in China, development of emission inventories at urban and regional levels, analysis of energy and environmental impacts of advanced vehicle technologies and new transportation fuels, and the evaluation of policies relating to China's transport sector. She has published more than 30 papers in internationally refereed journals.

Dr. Chunbo MA is Assistant Professor in the School of Agriculture and Resource Economics and Centre for Environmental Economics and Policy at the University of Western Australia. He was recently awarded an Australia-China Emerging Future Leaders Fellowship for low-emissions coal technology by the Australian Academy of Technological Science and Engineering. His research interests lie primarily in renewable energy markets and policy analysis, the Australia-China electricity market, environmental consumerism, non-market valuation and China's energy economy.

Dr. Shengwei WANG is Professor in the Department of Building Services Engineering and Director of the Research Laboratory of Building Energy and Automation at Hong Kong Polytechnic University. His research interests include the application of building energy management, building/HVAC diagnosis and optimal control, and intelligent building and building automation technology. He has conducted over 25 projects funded by government agencies and other organisations, and has published six books and over 300 papers in internationally refereed journals.

B. Su and E. Thomson (eds.), *China's Energy Efficiency and Conservation*,
SpringerBriefs in Environment, Security, Development and Peace 30,
DOI 10.1007/978-981-10-0737-8

Dr. Libo WU is Professor in the Energy and Environmental Economics Department, Executive Director of the Center for Energy Economics and Strategies Studies, and Deputy Director of the Center for Environmental Economics at Fudan University, China. She has published more than 30 papers in international and Chinese-refereed journals, and conducted dozens of projects funded by the local government in Shanghai and national agencies in China. She was awarded the "Excellent Achievement of National Energy Strategic Studies" in 2012 by China's National Department of Energy.

Dr. Jiahai YUAN is Associate Professor in the School of Economics and Management at North China Electric Power University, China. He was a visiting scholar at the Erb Institute for Global Sustainable Enterprises at the University of Michigan from 2011–12. His research interests focus on energy economics, energy planning and energy policy, with a special interest in China's energy issues. He has published around 40 papers in international and Chinese-refereed journals.

About this Book

China, with the largest markets and manufacturing factories in the world, plays an important role in global energy use and carbon emissions. China has made remarkable economic progress over the last decade, resulting in a sharp increase in its overall energy use and GHG emissions. In 2007, China overtook the United States as the world's largest emitter of CO_2 emissions. In 2010, China became the world's largest energy consumer. Many countries, international organisations and research institutes have realised the importance of understanding China's energy policies, and thus a large number of studies on China's energy and emissions issues have emerged over the last decade.

In China's 12th Five-Year Plan (FYP), the central government set a 16 % reduction target for national energy intensity and a 17 % reduction target for carbon intensity. On 25 January 2013, China's State Council further announced the "12th Five-Year Plan for Energy Development". The plan summarises the country's major achievements in energy development in the 11th FYP period (2006–10), and clarifies national energy policy priorities through an ambitious set of infrastructure and market targets in the 12th FYP period (2011–15). It also serves as a useful basis to benchmark China's progress in energy.

This officially released energy development plan not only identifies various aspects of the energy challenges faced by China's central/local governments, but also provides an opportunity to study how best to achieve green growth and a low-carbon transition in a developing country like China. The progress of China's carbon mitigation policies also has significant impacts on the ongoing international climate change negotiations. Therefore, both policy-makers and decision-makers in China and other countries can benefit from studying the challenges and opportunities in China's energy development.

More information on this book is at: http://www.afes-press-books.de/html/SpringerBriefs_ESDP_30-31.htm.

B. Su and E. Thomson (eds.), *China's Energy Efficiency and Conservation*,
SpringerBriefs in Environment, Security, Development and Peace 30,
DOI 10.1007/978-981-10-0737-8

MIX
Papier aus verantwortungsvollen Quellen
Paper from responsible sources
FSC® C105338

If you have any concerns about our products,
you can contact us on
ProductSafety@springernature.com

In case Publisher is established outside the EU,
the EU authorized representative is:
Springer Nature Customer Service Center GmbH
Europaplatz 3, 69115 Heidelberg, Germany

Printed by Libri Plureos GmbH
in Hamburg, Germany